DARWIN'S LOST WORLD

DARWIN'S LOST WORLD

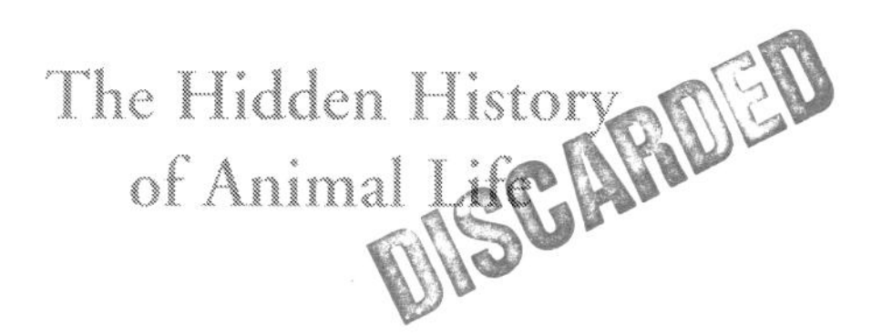

The Hidden History of Animal Life

MARTIN BRASIER

OXFORD
UNIVERSITY PRESS

OXFORD
UNIVERSITY PRESS

Great Clarendon Street, Oxford OX2 6DP

Oxford University Press is a department of the University of Oxford.
It furthers the University's objective of excellence in research, scholarship,
and education by publishing worldwide in

Oxford New York

Auckland Cape Town Dar es Salaam Hong Kong Karachi
Kuala Lumpur Madrid Melbourne Mexico City Nairobi
New Delhi Shanghai Taipei Toronto

With offices in

Argentina Austria Brazil Chile Czech Republic France Greece
Guatemala Hungary Italy Japan Poland Portugal Singapore
South Korea Switzerland Thailand Turkey Ukraine Vietnam

Oxford is a registered trade mark of Oxford University Press
in the UK and in certain other countries

Published in the United States
by Oxford University Press Inc., New York

First published 2009

British Library Cataloguing in Publication Data
Data available

Library of Congress Cataloging in Publication Data
Data available

Typeset by SPI Publisher Services, Pondicherry, India
Printed in the UK
on acid-free paper by
CPI William Clowes Beccles NR34 7TL

ISBN 978-0-19-954897-2

1 3 5 7 9 10 8 6 4 2

CONTENTS

PREFACE

Some 150 years ago, in 1859, Charles Darwin was greatly puzzled by a seeming absence of animal fossils in rocks older than the Cambrian period. He drew attention to a veritable Lost World that was later found to have spanned more than eighty per cent of Earth history. This book tells the story of his lost world, and of the quest to rescue its hidden history from the fossil record.

Intriguingly, such a quest did not really begin until 1958, some hundred years after Darwin. Why did an understanding take so long? Arguably it was because it was, and still remains, a very big and very difficult problem. Its study now involves the whole of the natural sciences. Progress has been a matter of slow attrition. For most of this time, for example, there has been no concept of the vast duration of Precambrian time, nor any evidence for a distinct biota.

This book follows the story of my own research history, beginning with a cruise as Ship's Naturalist on *HMS Fawn* studying Caribbean marine ecosystems. Like my own researches, it then pushes ever further backwards through time, from an inquisition into the nature of the Cambrian explosion and the enigmatic Ediacara biota some 600 to 500 million years ago, towards the emergence of the earliest complex cells some 1000 million years back. Each step backwards in time has drawn me towards ever more remote and little known parts of the planetary landscape, and towards equally puzzling parts of the human mental landscape. I have therefore sought, in each chapter, to put some of the major questions into context by descriptions of premier field locations from around the world, enlivened by descriptions of their fossils, their fossil hunters, and their puzzles.

My hope is that the book will show just how rich and diverse have been our ways of thinking about the earliest life forms, written in words that can hopefully be read with ease and enjoyment. Good science is, after all, not just about facts. It should be a form of play. If a thing is not playful, it is probably not good science. Each generation has therefore come up with its own favourite solution to the question—*whence cometh life*?—only to watch it fall as the next generation of science and scientists has arrived on the scene and found even better solutions. Deep down, my hope is that the book will show how my subject works as a science, how the questions are being shaped, and how the early fossil record of animal life may yet be decoded, bringing the world of ancient and modern life right to the doorstep of all those who are curious and wish to learn about the rich history of life beneath their feet.

Here, then, is your passport to becoming a Time Traveller, and to making your own exciting discoveries about the world in which we really live. The fossil record is your best guide for decoding pattern and process and the meaning of life. And the starting point for the reading of patterns is your own natural curiosity spiced with a modicum of *doubt*. Happily, science is a uniquely valuable system for the measurement of doubt.

As my colleague Andy Knoll at Harvard has so rightly put it: 'science is a richly social endeavour.' Nothing in science would be possible without the support of a network of friends and colleagues. I here express my deep gratitude to the following fine teachers and mentors for encouraging my involvement in a lifetime's research into early life, roughly from the 1960s onward: John Dewey, Tony Barber, Bill Smith, Martin Glaessner, Perce Allen, Roland Goldring, Stewart McKerrow, Françoise and Max Debrenne, Michael House, John Cowie, Peter Cook, John Shergold, and Stephen Moorbath. I thank my colleagues Jonathan Antcliffe and Latha Menon for acting as major catalysts for this popular (or possibly unpopular) account and for giving me greatly appreciated advice upon the written word. The following geologists from around the world are here lauded for their invaluable assistance during field work, often in remote and hostile places, followed up by laboratory work, over four decades: Owen Green for his kind support in field and lab, and for running the Palaeobiology Labs at Oxford; my extended family of students and protégés including Duncan McIlroy, Graham Shields, Louise Purton, Gretta McCarron, David Wacey, Jonathan Leather, Zhou Chuanming, Nicola McLoughlin, Jon Antcliffe, Maia Schweizer, Richard Callow, Alexander Liu, Leila Battison, and Latha Menon for so many years of lively and fruitful discussion in field, lab and pub; and the following for their much valued support and friendship in the field: Alexei Rozanov, Andrei Zhuravlev, and Vsevelod Khomentovsky in Siberia and Mongolia from 1990 to 1993; Xiang Liwen, Xing Yusheng, Yue Zhao, Jiang Zhiwen, Luo Huilin, He Tinggui, and Sun Weiguo in China from 1986 to 2007; Lena Zhegallo, D. Dorjnamjaa, Bat-Ireedui, Rachael Wood, Simon Conway Morris, and Stefan Bengtson in Mongolia from 1991 to 1993; Dhiraj Banerjee in India in 1990 and Pratap Singh for the donation of key Tal material; Joachim Amthor, Salim El Maskery, Philip Allen, and John Grotzinger in Oman; Petroleum Development Oman and the staff of Shell International for their generous support of Oman field work from 1994 to 2000; Philip Allen and John Grotzinger for showing me how to be a Precambrian sedimentologist; Bahaeddin Hamdi for generous provision of Iran samples and field notes; Eladio Liñan, Antonio Perejon, and Miguel-Angel de San José in Spain from 1978; Trevor Ford, Helen Boynton, Mike Harrison, Lady Martin, Mike Howe, John Carney, and numerous other kind people for their support of fieldwork in England and Wales at various times since 1965; Michael and Alison Lewis for making their Welsh farmhouse a home to Cambrian fieldwork over many decades; Mike Anderson, Ed Landing, Guy Narbonne, Caas van Staal, Bob Dalrymple, and Duncan McIlroy for major help with fieldwork in Newfoundland and Nova Scotia since 1987; John Hanchar and his team at Memorial University in Newoundland for their encouragement and support; John Lindsay of NASA and Cris Stoakes in Perth for their enormous enthusiasm and encouragement and greatly valued companionship and hospitality during fieldwork across Australia from 1998 to 2006; Jim Gehling, Dave McKirdy, Richard Jenkins, and Pierre Kruse for their guidance and hospitality in the Adelaide and Darwin regions of Australia in 1998. There are, alas, very many other friends, colleagues and early teachers not mentioned by name in this list, and I sincerely hope they will accept my thanks for their undoubted contributions here.

Last of all, I wish to thank my dear, and now departed, Mum and Dad for helping to imbue in me a great love for nature, the planet and the past beneath my feet; my wife and dear companion Cecilia for her unstinting help with fieldwork in Sweden, Norway, Spain, England, Wales, and Scotland over three decades, and for being a perfect 'geology widow' for several months of each year; and our children Matthew, Alex, and Zoe for their lively wit and enthusiasm during those many holidays spent on the rocks in Islay, Jura, Assynt, and the Dordogne. This book would not have been possible without their constant support and encouragement.

Oxford
July 2008
M.D.B.

LIST OF FIGURES

All drawings are by the author unless otherwise mentioned.

LIST OF COLOUR PLATES

CHAPTER ONE

IN SEARCH OF LOST WORLDS

Darwin's Great Dilemma

It is January, 1859. Imagine being a visitor, seated on a sofa in Darwin's large and dark study at Down House in Kent. The Christmas decorations have been put away. The smell of leather-covered books, gas lamps and moth balls rises up from the walls of this well-used room. Over the past few months, our world-weary naturalist has been drawing together the last remaining thoughts for his new work, to be called 'On the Origin of Species'. Its contents have been gathered from notes and observations made since 1831, an epic of nearly thirty years' gestation.

At first, we see him sitting in his easy chair, scratching away eagerly upon a board resting on his knee. We expect him to look engrossed and satisfied, and for a while he does. But suddenly, he looks up and scowls and then starts to pace nervously around the room, tapping the palm of his hand with a pen. After a minute, he stops to pick up a fossil trilobite from the mantle shelf. It is one of the oldest animal fossils known from the geological record. Looking like a little woodlouse trapped within layers of black slate, it is without any eyes—a completely blind trilobite. Teasingly called *Agnostus*, its name could be taken to mean 'without a knowledge of God'—blind to the Creator. Darwin turns this offending fossil over and over, as though searching for something. He then lets out three words, with a hint of genteel exasperation: 'Inexplicable... absolutely inexplicable.' With that, he sits down to write this phrase on his notepad: '*The case at present must remain inexplicable; and may be truly urged as a valid argument against my views here entertained.*'[1]

Our story begins here, with this great puzzle set in 1859 (see Plate 1). Charles Darwin clearly disliked the mad rush to publish 'On the Origin of Species'. It was full of risks to his reputation, to his health, and to his peace of mind. He had spent many a restless night, worrying over the shape of the arguments in each chapter, checking and rechecking the language to ensure that all the sentences sounded authoritative, measured, and balanced. Like many a scientist driven unwillingly towards the arena of public debate by the adrenalin of a new discovery, he had no doubt been hearing the voices of his enemies hissing at him, like gas lamps in the quiet of the Victorian evening. And by 1859, Darwin had made a rather splendid enemy. His nemesis was the eminent scientist, Richard Owen, a highly intelligent, outspoken, ambitious, and more than usually unpleasant Victorian anatomist. In appearance, Owen was dark, dapper, striking and to our modern minds, perhaps, rather sinister-looking—a kind of Professor Moriarty from the Sherlock Holmes stories.[2] More to the point, Owen boasted a brace of good connections. He was Superintendent of Natural History at the British Museum, a Member of the Athenaeum Club, and close to the bosom of the Royal Family. He was also famously arrogant, taking great pleasure from sneering at intellectual competitors such as Darwin. The latter had written bitterly to a friend that Owen was being 'Spiteful... extremely malignant, clever, and... damaging' towards him.[3]

But publication of the *Origin of Species* by Darwin, however distasteful, had become unavoidable, ever since Alfred Russell Wallace had written to him from the jungles of the Moluccas. Both naturalists had, quite independently, stumbled upon the same, dangerous and earth-shattering, conclusion: that natural selection causes new species to form and life to evolve with time. And that this simple process, operating alone in nature, is the Rosetta Stone that helps to explain most of the diversity of life, both present and past.[4]

To win the argument on evolution back in 1859, Darwin needed to point not only to a plausible pattern but to an ultimate 'first cause' for evolution. He had indeed stumbled upon a stunningly simple 'first cause', natural selection, that does away with supernatural involvement in the diversification of life 'from simple beginnings'. Like a flock of fledglings flying home across a storm-tossed sea, Darwin had observed that only the strongest and fittest within a living population will survive to the end of their journey. He had identified that life was a race against endlessly winnowing forces, like flying against the rain, the wind, and the waves.[5] But he could not explain why these populations varied—nor, indeed, how such variation was transmitted from one generation to the next.

The flock of young birds is merely our metaphor, of course. What was needed by Darwin was something more concrete—an appropriate set of biological experiments, something easy to study and close at hand. Something from the dinner table, perhaps. Interestingly, he decided to settle upon viands from the Sunday roast for tackling this question. But he had inadvertently chosen the wrong dish, working upon fancy homing pigeons rather than upon garden peas. A Moravian monk named Gregor Mendel was shortly about to discover the rules of genetics from his experiments on pea plants over a series of years in the monastery gardens.[6] Unfortunately

Darwin knew nothing of this work. And quite a lot was still worrying him—he had found evidence for the workings of natural selection, but could not explain the mechanism of inheritance. Nor could Darwin yet point to any detailed evidence for evolution in the fossil record.

As if these were not problems enough, the greatest puzzle facing Darwin was actually rather shocking. An inexplicable surprise was just beginning to emerge from within the geological record itself: the greater part of the rock record appeared to preserve no physical evidence for life on Earth at all.[7] That is to say, no geologist back in 1859 was able to point to any convincing kinds of fossil in the most ancient rocks on Earth, which today we call the Precambrian rocks. There were no clear animal fossils below the trilobites. Older rocks remained oddly silent. That would not matter much if the Precambrian was only a short period of time. But, as we shall shortly see, Darwin knew that this 'silence' was no brief aberration. It had spanned the greater part of Earth history.[8]

Just how long ago all this took place, or rather didn't take place, also became part of a nightmare for Charles Darwin. In early editions of the *Origin*, he had implicitly been thinking of many hundreds of millions of years ago. But by the sixth edition of 1872, he obviously felt a bit rattled:

> Here we encounter a formidable objection; for it seems doubtful whether the earth, in a fit state for the habitation of living creatures, has lasted long enough. Sir W. Thompson [Lord Kelvin] concludes that the consolidation of the crust can hardly have occurred less than 20 or more than 400 million years ago, but probably not less than 98 or more than 200 million years. These very wide limits show how doubtful the data are; and other elements may have hereafter to be introduced to the problem. Mr Croll estimates that about 60 million years have

> elapsed since the Cambrian period but this, judging from the small amount of change since the commencement of the Glacial epoch, appears a very short time for the many and great mutations of life, which have certainly occurred since the Cambrian formation; and the previous 140 million years can hardly be considered as sufficient for the development of the varied forms of life which already existed during the Cambrian period.[9]

Fortunately, both Mr Croll and Sir W. Thompson were to prove very wide of the mark indeed. Thanks to the twentieth-century discovery of radiogenic isotopes, which can be used to date rocks accurately, we now know that Precambrian rocks must have been laid down during the first 80 per cent or so of all of Earth history, from about 4560 to 542 million years ago. Cambrian and younger rocks, with all their fossils—from trilobites and ammonites to dinosaurs and ape men—therefore provide little more than a footnote to the history of our planet. When Victorian geologists crossed over that threshold which we now call the Precambrian–Cambrian boundary,[10] nearly everything appeared to change. Not least among these revolutions was the astonishing observation that nearly all major animal groups appear rapidly in the fossil record, within just a few tens of metres of rock, or just a few million years. That is to say, complex animal life seemingly appeared almost 'overnight' in geological terms. This paradox, of a long period without known life (then called the Azoic) followed by a rapid revelation of fossils (now called the Phanerozoic), must have felt like a cruel challenge to Darwin in 1859. Only the maddest of French Republican scientists would have dared to contemplate this as a bloody revolution *within* the history of life. It was so terribly... un-English.

Darwin was therefore forced to concede that this seemingly abrupt appearance of complex animal life near the beginning of the Cambrian, now called the Cambrian explosion, could be seen as a major stumbling block for his evolutionary theory—a mystery that some have called 'Darwin's Dilemma'.[11] Worse still, the puzzle of Darwin's missing fossils could be seen as evidence for the act of Creation itself. That was Sir Roderick Murchison's view, although his mentor Charles Lyell was trying to keep an open mind on the matter. Darwin was therefore careful to speak in very cautious terms within the *Origin of Species* about the absence of any fossil ancestors or obvious intermediates between the known animal groups. He postulated a 'lost world' too dim to make out through the mists of time:

> There is another and allied difficulty, which is much graver. I allude to the manner in which numbers of species of the same group, suddenly appear in the lowest known fossiliferous rocks... I cannot doubt that all Silurian trilobites[12] have descended from some one crustacean, which must have lived long before the Silurian age, and which probably differed greatly from any known animal[13]... Consequently, if my theory be true, it is indisputable that before the lowest Silurian stratum was deposited, long periods elapsed, as long, or probably far longer than, the whole interval from the Silurian age to the present day; and that during these vast, yet quite unknown, periods of time, the world swarmed with living creatures.[14]

Evidence from a living Lost World

Darwin believed that life had existed in periods long before the Cambrian, and that fossil evidence for this would eventually be found. On *HMS Beagle*, he had to content himself with clues about the early history of life gathered from the islands of the Galapagos. For many a budding scientist in later times, their dream was to discover another Lost World, revealing the deeper history of Life. It was certainly mine, ever since picking up a copy of Sir Arthur Conan Doyle's adventure story of the same name, about a scientific expedition that set out to discover long-lost life forms living on a plateau hidden deep within the jungles of South America.[15] And in 1970, I was to have the chance to explore for myself just such a lost world. Shortly after graduating, and to my considerable surprise, I found myself as Ship's Naturalist, lying in a bunk aboard *HMS Fawn*, sailing out of Devonport dockyard behind *HMS Fox*, and bound for the Caribbean.[16] The *Fawn* was a sleek white surveying ship of 1160 tons displacement, with a buff coloured funnel and gleaming teak decks. Stirringly, she was also a successor ship to *HMS Beagle*. When I joined her in the Naval Dockyards at Devonport, she and her sister ships (*HMS Fox, Beagle*, and *Bulldog*) were the pride of the Hydrographic Division of the Royal Navy.[17] *Fawn* even looked like a millionaire's yacht—especially when at anchor in a Cayman lagoon on a moonlit night.

Our brief was to chart reefs and lagoons, and to gather together an environmental rollcall of marine life from this unspoilt portion of Paradise. In particular, our plan was to make detailed charts of two great natural hazards to shipping in this dwindling pond of the British Empire. One of these was a huge island that no one alive had ever seen, called Pedro Bank. For most of the geologically recent past, Pedro Bank had been an island as big and as lush as Jamaica. But nobody has ever seen it because it sank beneath the waves some ten thousand years ago, at the end of the last Ice Age, like the legendary Atlantis. The second hazard was an even larger lost island, called the Barbuda Bank. This lay just to the north of Antigua, where Nelson had his harbour, as for a while did we.

The sea, the sky, the bone-white beach and zesty afternoon breeze lightened our daily chores of sampling and echo-sounding on board ship. Long spells at sea were punctuated by lively visits to friendly island states scattered between the Orinoco and Florida. For one month, we were even sent on a mission to clear the Bahama Banks of pirates. President Fidel Castro had complained to the British Government that the north coast of Cuba was being raided by pirates who were holed-up in some remote islands of the Bahamas. Our governmental response was to send in the Royal Navy.

Suddenly, we had to abandon the making of charts and the measuring of dainty sea shells, to take up rifle practice. As night fell, we would scan the radar for signs of unexpected vessels. One night, at last the cry went out—a suspicious object had indeed been spied, floating in the water about one hundred yards to starboard. Off went the search party to tackle this dark menace: illegal lobster pots, dozens of them, filled with tasty crayfish. We dined royally for a week. No pirates were ever seen of course—we were careful to make far too much noise for that.

The Bearded Lady

I had been at sea on board HMS *Fawn* for five months before my Galapagos moment arrived, in August 1970. My 'Galapagos' was to be Barbuda, at that time one of the most unspoilt islands in the tropical Atlantic.[18] This name will make any Spanish speaker smile, because it means, quite literally, 'the bearded lady', perhaps in reference to its jutting goatee of storm beaches. Christopher Columbus never saw Barbuda because it is so very low lying and easily concealed behind the Caribbean swell. Indeed, this invisibility makes it one of the greatest navigational hazards in the region. Later seafarers also largely ignored Barbuda because its soils are poor and thin, and its climate distinctly arid. There is some vegetation, of course, but much of it is little more than a wilderness of Turk's Head cactus, sword-leaved *Agave*, manchineel, and mangrove. The island has little farming, and only a thousand souls then lived in its single settlement called Codrington, making a modest living from conch and crayfish in the surrounding reefs and lagoons.

My own dream during the cruise of *HMS Fawn* was to see how the evolutionary and environmental history of modern reefs and lagoons might be traced backwards in time, using proxies from the fossil record. Was that even remotely possible? That August, a quick encirclement of the island on foot and horseback showed that four main features make up the primeval scenery of Barbuda: the highlands and the lowlands, the lagoons and the reefs (see Figure 1).

First, and most ancient, is that remote plateau of limestone—called the Highlands—which rises up sharply, 100 feet or so, above the jungle. When I first set foot on the island, there was no proper road out from the little village of Codrington towards this mysterious plateau, which was then so densely vegetated that it was almost impassable without a cutlass. Seeing this for the first time was, indeed, like having a private invitation to explore Conan Doyle's Lost World.

These Barbudan Highlands are surrounded by a diadem of salt lakes and lagoons, strung out along the western or leeward side of the island—over a dozen of them, some active, some ancient, and each of them home to a distinct menagerie of plants and animals. The greatest of these salty lakes is called Codrington Lagoon, which is some three kilometres wide and ten or so kilometres long, and connected with the open ocean to the north by means of a long and winding tidal channel. Branching off this main body of water, we found a series of smaller lagoons. Each of these smaller lagoons is separated from its neighbour by a narrow beach of pink

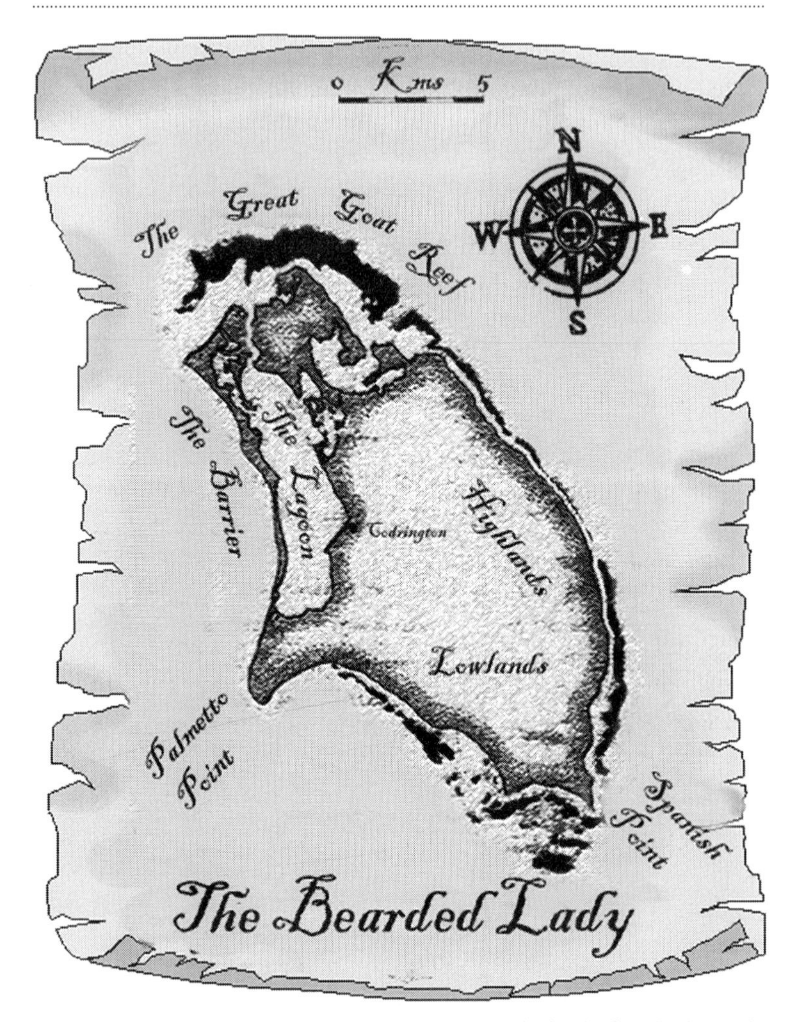

Figure 1. A Living Lost World. Map of the tropical island of Barbuda, in the eastern Caribbean, where the author began his researches into Darwin's Lost World. The Caribbean Sea lies to the west, and the main areas of coral reef (shaded in black) face the Atlantic Ocean to the east.

and white sea shells, often with its own distinctive character. There would be mangrove and frigate birds on one ridge, coco palms and giant mosquitoes on another, and so on. Most days, the only craft we would see while mapping this watery wilderness was a schooner sailing down the main lagoon, bringing the much needed weekly rations from Antigua.

Around Barbuda lie three reef tracts, each sheltering the island from the annual tropical storms and their steaming white breakers. These reef tracts were then some of the finest in the whole of the tropical Atlantic; vigorous because of the huge oceanic swell that pounds across them. Indeed, there is almost no part of the island where you could not hear the distant boom of waves thundering across the reefs. A hair-raising roar would therefore assault us as we approached the reefs for mapping each morning. The main fringing reef of the island is some fifteen kilometres long and clings to the rocky shoreline along much of the eastern seaboard. To the north and south of the island, the fringing reef fans out into luxuriant reef gardens, bathed in waters as warm and clear as a baby's bath. These coral reefs conceal a darker harvest, however—a scattering of some 200 shipwrecks. But both algae and corals delight in shipwrecks—they use them to build upwards and outwards along the tract.

It was tempting to share the rapture felt by young Charles Darwin as I escaped the claustrophobia of *HMS Fawn*, after many months spent at sea with her strict routines, stuffy protocols and endless meals of boiled cabbage, and sailed away in a small boat towards my tropical island. But whereas the young Charles Darwin spotted oddities in birds and land turtles across his islands, that is to say, in terms of *space*, I was, in my modest way—and I had much to be modest about—preparing to stumble upon some oddities in terms of *time*.

In a grain of sand

During the last quarter of 1970, we mapped the distribution of microbes, plants, and animals around the island. Days would typically begin at dawn, speeding with companions across the lagoon in our motor boat, followed by snorkelling and sampling in some hitherto unexplored part of this watery Eden.[19] Work would continue in the boat until mid afternoon, when the trade wind got too strong for comfort. We would then make our way back towards the little jetty in the south-east corner of Codrington Lagoon, to beach the boat before brewing up for afternoon tea.

Back in the village, we had set up a little laboratory within an abandoned cotton mill—called the Ginnery—right next to the lagoon. In these hot and windy afternoons onshore, we would pack our sample jars, full of pink sand and gaudy shells, in readiness for shipment back home. In the gloom of the evening, a hurricane lamp provided a modicum of light for us to find our way around inside the dingy warehouse. The tinkle of Gladwin Nedd's Steel Band would then waft through the evening air from the Timbuck-One Saloon, and we would watch our lamp being dive-bombed by a squadron of moths, mantids, and beetles. Not that we were safe from arthropod attack during the day. More venomous creatures were on the prowl around the Ginnery than we could ever hope to see. One morning for example, on getting out of my camp bed, I jumped into my shorts a little too hastily. Just as I was pulling them on, I spotted a fluffy Tarantula spider nestling comfortably in the gusset. We kept 'Tara' in a jar for months. But while tarantulas and giant centipedes seemed to be crawling everywhere, we gradually learned to keep them out of the shower and, most importantly, from under the toilet seat.

It was within this little zoo that was I able to set up a small binocular microscope and study living and fossil shells from around the island. The gathering and identifying of plants, sponges, corals, and seashells began to mount as the weeks turned into months. By late September, we had collected and mapped dozens of different types of organism across the floor of Codrington Lagoon.[20] A stunning surprise, though, was the sheer abundance and variety of life that became visible when I started to look down the barrel of my microscope. The diversity of little shells seemed to multiply logarithmically every time the level of magnification was raised. For example, a square metre of lagoon or reef would visibly support a dozen or so species of macroscopic mollusc shells. But there was vastly more variety and abundance when the lens was racked down to shells just a millimetre or so across. And another huge increase in diversity emerged when I scanned the microscope down to creatures less than a tenth of a millimetre wide, many of them of exquisite beauty (see Plate 2). One thing was beginning to become clear: there is a 'fractal' quality to the biosphere.

When I got back to the old Cotton Ginnery each afternoon, I would set about 'seeing this little world in a grain of sand'. One of the first things to be tested was the number of species present within a 'teaspoon' of sand and mud from Goat Reef. In a sample of seafloor, as bleached and unremarkable as a patch of sand from the Arabian desert, almost ten thousand individuals and a hundred species of foraminiferid protozoans were present. And this count did not take into account any of the other tiny creatures or the nearly invisible microbes. At that time, little could really have prepared me for this surprise, but something similar is now known to hold true across the whole of the natural world. It is called the biological scaling law: there are lots of little creatures and progressively fewer large ones. If I was going to trace the history of life back into deep time, it would clearly be wise to forget about dinosaurs. It was the smaller creatures I would need to watch.

On Goat Reef

Looking east from the beach towards the reef tract of Barbuda, the great Atlantic rollers appear as a long white ridge that dances up and down on the skyline. These waves are symptoms of the trade winds that refresh this part of the tropics. Under such conditions, the coral formations are sculpted into buttresses and shelves, corridors and caverns of all shapes and sizes. Some parts of the reef are to be found extending their branches up into the sunshine while others lurk in self-made gloom. Each part of the reef therefore has its own biota, with light-loving corals and algae thriving on the sunlit tops, and shade-loving sponges and protozoa nestling down in the dim recesses. One morning when we brought pieces of this living rock up into the boat, I was rather surprised to find it full of holes—circles and slits, tubes and tunnels of all shapes and sizes. These caverns are entirely natural, of course, providing homes for a rich menagerie of invertebrates, such as yellow sponges, pink sea squirts and purple brittle stars. The more holey the habitat, it seemed, the richer was the tally of life. There was truth, after all, in that old saying: life is a search to maintain surface area.

The chief architect of this reefal framework is a coral called *Acropora*, the elk horn coral. In fact each elk horn is built by a colony of genetically identical coral polyps, natural clones that form colonies which may be some hundreds of years old. Diving around elk horn coral can be a dangerous and painful enterprise:

not only can the tiny corallites inflict painful rashes in the epidermis, much like a jellyfish sting, but their hard and chalky constructions will shred human skin like a knife, given half a chance. It does well to remember that elk horn corals have torn out the bowels of many a ship, allowing the reef to dine at night upon sailors.

Biologists have for long realized that corals like *Acropora* are cnidarians—cup-shaped animals with a flower-like radial symmetry. All cnidarians are provided with a simple but very effective survival kit: tentacles with stinging cells plus a seemingly unhygienic but effective gut: there is no anus. The coral has therefore to eject its waste through its single opening, called the mouth. Hence, in terms of the tree of life, cnidarians are regarded as relatively primitive animals. Indeed, only the sponges are thought to branch lower down on the tree.[21] Another remarkable feature about reef-building corals is the way that they feed. Each tentacle is equipped with tiny stinging cells that are primed to harpoon small animals as they pass. Corals are particularly adept at catching zooplankton, tiny crustaceans and animal larvae that lurk in the depths during the day but come up to feed in the water column at night.

The inner layers of coral polyps along the reef crest can also appear like a commercial greenhouse, with rows of simple plant-like cells that have been enticed from the water column. In the natural state, these plant-like cells show flame red pigments and bear whip-like threads that make them twist and turn in the ocean like whirling dervishes. Hence their name 'terrible whips' or, more correctly, dinoflagellates. This group, when let loose on their own, can cause the equally alarming 'red tides', which not only poison commercial shell fish with toxins but also cause mass mortality of fish and sea birds.

Dinoflagellates are attracted to a life within corals by a seductive offer of *Free Fertilizer*! But they are then obliged to repay this debt, day-in day-out, by gratefully churning out food for the coral with the aid of captured sunbeams. Both creatures benefit from this symbiosis, it is true.[22] It is a little hard, though, to say whether the relationship is like a marriage, or like slavery. My suspicion is the latter. Such sweated labour underpins the ecology of the whole coral reef, which has become dependent upon a trickle-up effect of nutrients passed along from symbionts to their hosts and thence to the next food layers.

The underside of elk horn skeletons are commonly devoid of living coral polyps. These bare patches provide valuable space for other creatures that are trying to make a living from the reef. Some of these will drill their subterranean dwellings into the old coral rock. Occasionally, they get a little too successful and cause the coral heads to collapse on to the seabed, where they will finally crumble into coral sand. Others have the opposite effect, of binding old coral skeletons together through the addition of new chalky layers. Coralline algae are famous for defending, in this way, the whole reef edifice against the ravages of time and tempest.

But one of these heroic reef binders is not an alga. It is a ruby red protozoan, a single-celled foraminiferid called *Homotrema rubrum*. This tiny beast also has a charming way of feeding. It extends jelly like threads, called pseudopodia, out of its little red shell into the surging seawater and, biding its time, it garners passing sponge spicules from the water column. Each spicule is shaped like a tiny glass needle. *Homotrema* can therefore use it as a kind of fishing rod, allowing its pseudopodia to stretch out into the warm water to entrap tasty food particles. Man may be a tool-maker. But even a protozoan can make good use of a toolkit.

Foraminifera like *Homotrema* are thought to lie below the base of the tree of animal life. That is because they are single-celled, with neither tissues nor organs of the kinds found in true animals. But they do share, with animals, a fancy for feeding on other living matter. In other words, *they like to eat things*, especially bacteria. And while they may be of rather lowly status, they make up for this by being astonishingly abundant. They can form as much as 90 per cent of deep sea floor biomass in polar waters, and can flourish as rock forming microbes from the top to bottom layers of the ocean water column. *Homotrema* is therefore like a partner in an illustrious family business. So successful is this little creature today, that its ruby red shells enrich the ivory white detritus of the reefal shoreline, glowing like little red jewels, to the delight of beachcombers.

Down the Emerald Lagoon

Climbing back into the boat and speeding back down the channel towards Codrington Lagoon, where the sheltered waters are seldom more than neck deep, it is easy to spot a progressive change that takes place in the seafloor. Reef-building corals begin to disappear from the sea bottom when we reach the tidal channel, presumably because the waters here can become far too salty, owing to evaporation on hot and windy afternoons. An expansive surface area for life is, however, maintained in other ways: by the fine-grained nature of the muddy sediment; by kilometre-long submarine banks carpeted with Turtle Grass and Neptune's Shaving Brush (a fibrous kind of green alga); and by tangled forests of mangrove roots.

Mangroves living close to the tidal channel are often colonized, below the water line, by strikingly coloured sponges. Their blue, purple, and orange colours provide an indication of their different kinds of microbial guest, and they are especially fond of nurturing blue-green microbes called cyanobacteria. In terms of shape, some of the sponges on the mangrove roots look like golf balls, while others resemble flower vases and even church organ pipes. The animal nature of the sponge group, properly called the Porifera, was for long debated because they can look and behave a bit like plants. Unlike the corals, for example, sponges have no gut or nerve network. They do not recoil when attacked by fish. Being little more than a colony of cells, sponges can also take almost any shape and will grow in any almost direction. But they are far from passive. In one infamous experiment conducted in the early 1900s, a living sponge was disaggregated by squeezing it through the very fine mesh of a lady's silk stocking—presumably with her permission, and hopefully when she was no longer wearing it. The sponge cells seemed quite unperturbed and were able to 'pull themselves together' on the other side of the silk stocking. (This experiment is not something to be tried at home on an earthworm, or even on a jellyfish, because the outcome would be unpleasant.) In other words, they behave much more like a colony of cells than do the cells of a jellyfish or the cells in our own bodies. I have even observed sponge colonies creeping about, by means of cellular migration, at a rate of a few millimetres a day. They can also put out long streamers of cellular tissue, either to colonize new areas or to kill unwanted neighbours. This they do, like Nero's mother, by means of slow-acting poisons. Perhaps the strangest feature of living sponges is that they have little regard for body symmetry, with inhalant and exhalant openings scattered willy-nilly over their surface. This lack of symmetry in sponges is usually regarded as a primitive feature. It contrasts sharply with the beautiful symmetry of creatures that occupy the next rung on the ladder of life, such as the jellyfish and the corals.

Even so, the body wall of a sponge has great beauty, as is perhaps familiar if you ever take a bath. Careful inspection shows that it is full of tiny holes through which the seawater is drawn by thousands of little cells. These choanocytes will typically wave their tiny whip-like flagella in the manner of cheery supporters at a football match, causing a one-way flow of sea water from the outside inwards. But so tiny are the pores through which this water must travel that they are unsuited for filtering anything much bigger than a bacterium in size—about a thousandth of a millimetre across. Particles larger than a few thousandths of a millimetre across will tend to clog the pores, with fatal effect. We chordates can cough and sweat, but sponges have no body mechanism for cleaning their clogged pores. In such a dangerous situation, sponges are faced with two options. The first is simple: to remain in areas of water provided with laminar flow, not turbulent flow, and with good water clarity. This is what confines many modern sponges to the deep sea floor. The second is more subtle: they hire a cleaning service such as that provided, quite willingly, by invertebrates such as shrimps and brittle stars. This is how many sponges are thought to cope with detritus in the reefs and lagoons of Barbuda.

There is one more organism that we need to take a look at here. Vivid green in colour, it forms great clumps along the floor of the channel and flourishes in mangrove-enclosed ponds. Wherever it thrives, the seafloor is covered with something that looks a bit like snow-white cornflakes. Indeed, it is these 'cornflakes' of chalky material that make up much of the seafloor here, and in parts of the coral reef as well. When one of these green clumps is brought to the surface, we can see that it looks like a tiny Prickly Pear

Cactus. There are, however, neither specialized tissues nor flowers because it is really a kind of alga in which the photosynthetic pigments and nuclei float around wantonly within a very large cell—called a syncytium. Interestingly, this creature, named *Halimeda*, shows us one of the ways in which higher, multicellular organisms could have evolved: by the subdivision of this syncytium into cells. But there is more. A snorkel around the lagoons in the cover of darkness shows that *Halimeda* changes colour at night, bleaching to a bone white colour. That is because the green photosynthetic pigments have sneaked away from the surface of the seaweed, to lie protected within little canals inside the calcareous 'cornflakes'. It does this every night, as though trying to protect its valuable photosynthetic pigments from the grazers of darkness that haunt the lagoon floor—sea urchins, snails, and fish. These grazers of darkness come out from their hiding places as soon as the barracuda and other daytime predators have gone to sleep in the mangrove. In other words, the chalky 'cornflakes' of *Halimeda* are a deterrent to grazers. In an earlier world without grazers and predators—such as Darwin's Lost World perhaps—these cornflakes may not have been needed.

Up Cuffy Creek

From these mangrove swamps, it is but a short swim towards the more restricted lagoons and saltponds or creeks, where the balmy waters are usually rather shallow, and often less than knee-deep. Wading barefoot across the creek floor, the soft mud can prove distinctly ticklish as it squirms between the toes. A faintly sulphurous aroma, a bit like that from boiled spinach at dinner time on *HMS Fawn*, also wafts up from the mud here. That is because these muds are rich in organic matter, interestingly like the old lagoons we shall later meet that yield the first signs of animal life. The salty waters of Cuffy Creek quickly tell where the deepest skin scratches are, too. Neither corals nor seagrass can withstand the stresses here. In these restricted creeks, the habitable niches for animals are in rather short supply.

The floors of these creeks are covered with a natural fabric that rather resembles a low quality Persian rug, all woven together using pastel-coloured silky filaments (see Plate 2). These threads are, however, embedded within large amounts of snot-like mucilage, a feature rarely met with in Persian rugs, even the very cheapest ones. This slimy rug is called a 'microbial mat' or a 'biofilm' by biologists because of the dominance of cyanobacteria and other kinds of filamentous microbes. To a geologist, such a rug is known as a 'stromatolite'—meaning a layered or bedded stone. Ancient examples can be cabbage-like, and may preserve the remains of the earliest known communities. But here, in Barbuda, these living 'stromatolites' are grazed flat by high-spired snails such as *Batillaria*. This is a tough little mollusc that will feed happily until the dry season comes, when it will seal off its aperture and snooze for a while. Snail shells provide a lifeline for other creatures too. Fronds of the bottle-brush alga, *Batophora*, little more than a centimetre long, attach to shells here because there is little else to cling on to. And, in turn, the high surface area of each little bottle-brush alga provides a niche for thousands of tiny protozoans, foraminifera with the delightful name of *Quinqueloculina*, meaning 'a dwelling with five tiny chambers'.

Even these havens are obliged to give way to a monotonous carpet of microbial mats in the hottest and saltiest ponds. Water here, if present at all, is rapidly soaked up beneath the mat. Walking across the rubbery surface, a well-placed footprint will often penetrate some centimetres below the surface to reveal, within the mat, striking bands of different colours. This zonation shows that microbial life survives in the salt ponds by diving down into the darkness. Winning microbes keep themselves aloof in the top zone while the losers become progressively stratified downwards through the sediment, forming a hierarchy that is seemingly stricter than humans in a Caribbean holiday resort. Sun-worshipping cyanobacteria such as *Oscillatoria* are like superstars that wish to be seen thriving in all the best lit spots. Beneath them lies a thin layer of microbes that has turned beetroot red, because of the presence of a sulphur bacterium called *Thiocapsa*, which has purple photosynthetic pigments. They cannot tolerate oxygen.

The rest of the 'layer cake' looks and smells a bit like a rubbish tip. This whiff of rotten eggs betrays the presence of a sulphate-reducing bacterium called *Desulfovibrio*. Feeding slowly on organic matter in the sediment, these microbes produce hydrogen sulphide as a toxic by-product that, happily, also helps keep unwanted competition away. Sulphate-reducing microbes are like the lowlife in a downtown nightclub—they can bear neither light nor oxygen.

Within a mere finger's breadth beneath the surface of the mat, we have therefore seen a transit from 'oxygen heaven to anoxic hell'. These changes have largely been brought about by those kinds of 'primitive' microbe that scientists call prokaryotes. Prokaryotes are primitive because their chromosomes are not held together in a nucleus, but drift about wantonly in the cell. Not only that, but they also lack those other useful gadgets that tend to be found in a true eukaryote cell. In this respect, we might say that eukaryote cells are like the Swiss Army Knives of the cellular world—they come provided for almost every eventuality, having chloroplasts for photosynthesis, mitochondria for energy storage, and cilia for locomotion. Prokaryotes have none of these 'gadgets'. But they make up for this lack of gadgetry by their stupendous ability to grow rapidly when the conditions are right. And conditions are nearly always right for at least one kind of prokaryote to flourish. Though seldom more than one-thousandth of a millimetre across, prokaryotes are truly the rulers of the world. They get everywhere. Without them, our lives would be seriously less amusing. We would probably starve to death before we suffocated, but it would be a close call.

The Great Chain of Being

One day on Barbuda, I amused myself by arranging seashells and flotsam along a tide strand into something like the Great Chain of Being. Pieces of microbial mat, with smelly bacteria and cyanobacteria were made to form the base of the chain; all of these were prokaryotes—single celled and without a nucleus. Above these on the beach I laid out a handful of ruby red *Homotrema* shells—to stand for single-celled protozoans with cell organelles like a nucleus. Higher along the chain were placed some yellow sponges—with no real symmetry and no organs—and then some white pieces of coral, to stand for cnidarians with organs but neither blood vessels, nor kidneys, nor brains. Above these were arranged such examples of the major animal groups as I could muster: a star fish (for echinoderms), a worm tube (for annelids), a pink Queen Conch shell (for molluscs), a land crab (for arthropods), and a seabird (for our own phylum, the chordates).

Something like this almost feudal ranking of living creatures—from sponges to humans at least—has been known and remarked upon by philosophers and writers on medicine for millennia. Going back to the archetypes of the Greek philosopher Aristotle

(384–322 BC), this Great Chain of Being was becoming central to thinking about the natural world by the mid to late 1700s. But the wider evolutionary significance of this spectrum in living organisms was not widely discussed in public during the Christian era, until the chance of being burned alive in the market place began to recede. Emboldened by the Enlightenment of the eighteenth century, Erasmus Darwin—the illustrious grandfather of Charles Darwin—felt it was safe to write the following big and beautiful thought in his treatise called *Zoonomia*, published as early as 1794:

> Would it be too bold to imagine, that in the great length of time since the earth began to exist, perhaps millions of ages before the commencement of the history of mankind, would it be too bold to imagine, that all warm-blooded animals had arisen from one living filament, which the first great cause endowed with animality, with the power of acquiring new parts, attended with new propensities, directed by irritations, sensations, volitions, and associations; and thus possessing the faculty of continuing to improve by its own inherent activity, and of delivering down these improvements by generation to its posterity, world without end?[23]

Had either Baron Cuvier of France, or Richard Owen, ever been served up a banquet of things to eat—ranging from microbes to birds—during the early nineteenth century, they would have thought each organism in the menu had been created by a master chef—the creator God—who seldom made more than slight variations upon the standard recipe.[24] If both shrimps and lobsters had been on the menu, these would have been seen as little more than variations upon the basic theme of a 'crustacean archetype'. Likewise for echinoderms, molluscs, and so forth. Each of these was centred on an ideal body plan—an archetype—and there seemed no prospect of any intermediates between them.[25]

But in 1809, Jean Baptiste Lamarck, then the lowly-sounding 'Keeper of Insects, Shells, and Worms' at the *Musée d'Histoire Naturelle* in Paris, was starting to explore a much more radical idea—that the Great Chain of Being was more like lunchtime in a works canteen, where people had sat down at different times to eat their way through a *set menu.*[26] Some animal groups had started early on, as simple cells (the boiled rice, perhaps) and had now reached the final course of the monkey brains (ourselves). But the dinner gong for each lineage was successively later, and many of the diners had only reached the stage of a shrimp or squid course. Some were only just tucking into the jellyfish, while others—like infants—were still supping on the boiled rice. It was evolution, but not as we know it.

There was a problem, however, with this idea about life and its evolution. As Richard Owen was later to point out, higher animals such as fish and ourselves do not go through *all* the stages of lower animal kingdom during their embryonic development from egg to adult.[27] No human child, for example, goes through the stage of being a mollusc. Instead, it seems that each member of an animal phylum starts out with a *set menu* (the embryo) that becomes increasingly *a la carte* as the individual grows and matures. In pointing out this phenomenon of divergence from the *set menu* during growth, Owen was actually pointing the way towards the explanatory power of the Great Tree of Life, as later revealed by Charles Darwin in 1859.

For Darwin, this pattern of branching was to be found at all levels, from the genealogy of individuals to the divergence of species and even to the divergence of the animal groups—the animal phyla themselves. As we have seen, though, Charles Darwin was presented with a surprise in the fossil record: all the major animal groups seemed to appear rather abruptly and fully formed.

While trying to understand the Origin of Species, he found himself confronted with the Origin of Animal Phyla, an altogether bigger puzzle. According to his understanding at that time, those missing animal ancestors should have taken a very long time to diverge from their single ancestor. And they should have left behind them a few intermediates between the animal phyla—missing links—in the fossil record. But not a trace of those ancestors or intermediates had been found below the Cambrian by 1859.

A molecular puzzle

It took about a century of research following Darwin's great book of 1859 before the magnitude of the jump from prokaryotes (such as cyanobacteria) to eukaryotes (such as *Homotrema* and ourselves) could be starkly revealed as the biggest dividing line in the whole of life. But in another twenty years, another revolution was about to take place, following the discovery of DNA and RNA sequencing techniques. Molecular sequencing of the living world has latterly changed our perspective on the Great Tree of Life in two ways that are important for our story. First, it has shown us that prokaryotic cells are vastly more diverse than can be told from their shape or size alone. A teaspoon full of forest soil, for example, can contain up to five thousand different kinds. All of them can, however, be divided into one of two main types: Eubacteria, or 'true Bacteria', like the cyanobacterial mats of Cuffy Creek. And Archaea, like the methane producers found in our own guts. According to these molecular studies, we may share more in common with the Archaea than with the Eubacteria. If so, it is from whiffy Archaea that our own ancestors are widely thought to have evolved.[28] A second important finding shows that all the animal groups, from sponges to fish and ourselves, cluster very close together near the crown of the Tree.[29]

There are, however, many things which molecular sequencing cannot yet tell us about evolution. Molecules cannot tell us much about creatures that have never been studied—either because they are no longer alive, or because they have not yet been discovered. Our best hope of seeing these lost worlds is, in the view of myself and many colleagues, the fossil record itself. Nor can molecules yet tell us anything reliable about rates of change in evolution. The principle of the molecular tree is simple. It notes that differences in the genetic code accumulate with time, and that these similarities and differences can be measured and used to reconstruct the Great Tree of Life. Sampled in this way, living organisms can only form the tip of a branch on the Great Tree. But the similarities between them can be used to infer, without seeing them, links between the branches that lie beneath the canopy.

The idea of a molecular clock builds upon this molecular tree. It assumes that that the greater the number of differences in a given piece of code in two descendants, the longer the time since they diverged in geological time. And it says that, if we can be confident about the approximate rate at which mutations occur, we can then estimate how long ago two living branches diverged in the tree of life. Molecular clocks are a neat idea, and they have provided some interesting insights into relatively recent periods in the history of life. But molecular clocks are themselves utterly dependent upon fossils to calibrate their time scales. It is important to accept that there has been much misunderstanding about all of this. These clocks compare the rates of gene substitution along a given string of genetic code through time, calibrating it against various well-known fossils from the rock record. Backward projections from the earliest known fossil examples and into 'Darwin's Lost World' must therefore enter into a dark age filled with uncertainty. Unhappily for molecular biology, it is now clear that rates of substitution can vary widely—between creatures that live long or die soon, or between populations that are large or small, and so forth. And geology has shown that populations before the Cambrian explosion are likely to have changed dramatically in ways we could never have anticipated until recently, as we shall see.

To comprehend the problem with molecular clocks, imagine that you are looking at a murder scene. The Palaeontological Police have found a body beneath a railway line. And the Molecular Horologist has been asked to calculate the date of the victim's birth, by *using a current railway timetable*, even though the crime took place long before the track was laid down. But no English judge would likely admit such evidence into a court of law.[30] Regrettably, therefore, molecular clocks cannot yet be admitted into our own court as evidence for dating the origin of animals. Only animal fossils themselves will do. When there are no animal fossils in the rocks, then the clocks can be no better than their assumptions. And assumptions, as we shall see, can sometimes be bigger than the things they seek to explain.

A wrong question

This brings us to a puzzling question. Can living organisms, like those in Barbuda, provide us with the keys to Darwin's Lost World? Consider, for example, the microbial mats of Barbuda, with their cyanobacterial mats thriving at the surface and their smelly sulphate-reducers at depth. Could such ecosystems preserve something of an ancient world from, say, a time before the evolution of grazing animals and atmospheric oxygen? Or consider the popular notion that sponges and cnidarians, like those we have met in the reefs and lagoons, resemble the ancestors of 'higher animals' including ourselves. Shouldn't we be looking for them in the rocks before the Cambrian?

My case here is that these living creatures are not likely to provide the answer to our big question. That is because they provide an answer to the *wrong question*: 'what do the simplest animals look like today?' rather than: 'what did the simplest animals look like way-back in time?' We can now see, for example, that there is a remarkable degree of connectedness between modern creatures. Oxygen from 'higher' plants is needed to make the sulphate on which sulphate-reducing bacteria depend. Climbing further up the Great Tree of Life, we see that many protozoans depend upon things that branch even higher—things like corals and snails or seagrass. On the next branch, we find that sponges can be peculiarly dependent upon the cleaning service provided by 'higher' animals like brittle stars. And they need more complex animals to stir up the microbes on which they feed. Sponges are therefore highly adapted to the world of worms, shrimps and brittle stars. The same caution can be applied to the idea that early oceans should have swarmed with jellyfish. Such animals are, today, provided with specialised stinging cells-called cnidocytes, as well as with suprisingly complex eyes. But these seem best equipped to capture animals that lie 'higher' on the tree of life. In a world without worms and shrimps, would jellyfish have had any use for such stinging cells? It seems to me doubtful.

All organisms of supposedly lowly status, like protozoans, sponges, and corals, now seem to be hugely dependent upon a world tuned to the presence of higher beings, from molds to molluscs, and now us. Before the Cambrian, the pattern of life of these simpler creatures may have worked in markedly different ways. In Baron Cuvier's Chain of Being every creature was believed created by God and given a particular status. Those species remained static through time until they were intentionally snuffed out. Moving forward to Darwin's Tree of Life, however, the modern biosphere is seen by us as the branching tips of a great tree, arising from organic evolution over several billions of years. Darwinian evolution has required a lot of physical change within some lineages, such as vertebrates, while others have seemingly experienced physical change to a much lesser extent, such as cyanobacteria. But while simpler organisms may appear to resemble their ancestors more than complex ones, all have evolved to fit in with the complex ecology of the modern world.

In other words, we must not expect to translate our modern world and its biology far backwards in time. Life in the early biosphere was probably very different from anything we see today. The world before the Cambrian may have been more like a distant planet. This is the mystery I here call Darwin's Lost World. And its clues lie sleeping in the fossil record.

Oddities in time

During October of 1970, with hammer in hand, I circled the island of Barbuda, somewhat shakily perched on the back of a nag, in search of rocky outcrops that might help to answer this deeper question: *just how good is the fossil record?* I then discovered something rather curious: that the ancient biological communities of Barbuda—microbial mats and seagrass, mangrove and reef—were all selectively preserved in rocks just a few tens of metres inland from their modern counterparts. In fact, the older the fossils, the higher they were found to lie on such benches above sea level.

Closer examination of the cliffs around Barbuda also showed another intriguing pattern, that strong filters act—both for and against—the entry of dead organisms into the fossil record. Shelly fossils made of a chalky material called calcite had the highest chance of preservation. Under the microscope, I could therefore see the remains of calcareous algae, foraminiferid tests, oysters, and the ossicles of sea urchins, all of them made of calcite. In rocks dating to the last interglacial, which lie about 6 metres above present sea level—and are some 125,000 years old—molluscs and corals could be preserved but their metastable aragonite shells were often full of holes. In the higher and older rock benches, up to a million years old, such fragile shells were dissolved or replaced. It was no surprise, therefore, to find that even more delicate soft tissues, such as leaves, roots, tendons and muscles were seldom preserved at all. Even so, a few small windows could be used to look upon their story. Seagrass communities, for example, had left behind ghostly signals in the fossilized shelly biota.[31] And, in rare instances, even mangrove leaves and seagrass roots could be found buried within peaty clays deep beneath the lagoon. In other words, evidence could be seen for a natural spectrum in 'fossilization potential', ranging from *common* for the skeletons of echinoderms and foraminifera on the one hand, to *rare* for animal tissues and flowering plants on the other. Now Charles Darwin had predicted something a little like this back in 1859 when he wrote: 'No organism wholly soft can be preserved.' Though as we shall discover later, this prediction would prove somewhat wide of the mark for an Earth before the evolution of animals.

Encrypted in stone

How real, then, was the Cambrian Explosion? To answer this and similar puzzles, we need to try to decode the early fossil record. But, as we shall see, some claims for cracking this 4-billion-year-long code have recently proved to be false readings, while others show great leaps of intellectual grandeur, on a par with Champollion's decipherment of the Rosetta Stone. Strange fossils do indeed share many daunting similarities with lost languages such as Egyptian hieroglyphs. Both are preserved as arcane geometries—strange markings, glyphs, in rock—and both largely conceal their deeper meaning from us. Both are also capable of false translation. In the case of hieroglyphs, we need only remember one haunting reading conjured up by the romantic poet Percy Bysshe Shelley back in 1817, from a colossal statue of Rameses II, preserved along the Nile near Luxor

> And on the pedestal the words appear:
> 'My name is Ozymandias, King of Kings:
> Look on my works ye Mighty and despair!'
> Nothing beside remains. Round the decay
> Of that colossal wreck, boundless and bare,
> The lone and level sands stretch far away.

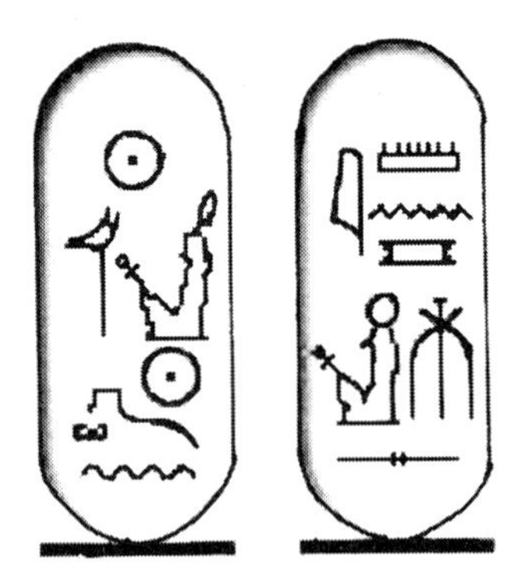

Shelley's reading is stirring and justly famous. But Shelley was not Champollion. No such words are actually present on the great statue. Decoding of the actual hieroglyphs of a cartouche on a nearby wall was impossible until Champollion got to work in 1822. This cartouche actually says 'User Maat Ra, Setep en Ra', which was the throne name of Rameses II (1293-1185 BC). It means 'The Justice of Ra is Powerful, Chosen of Ra.' Nothing about 'Ozymandias' here, nor about 'despair'. Shelley's reading was poetic, but it was sadly false.

To read the fossil runes correctly, it is necessary to adopt the mentality and techniques of the code breaker and the spy. First, we need to carefully record and decode the patterns. And then the patterns must be interpreted in terms of process. For example, deciphered hieroglyphs have allowed us to understand the challenges faced by past civilizations—of famine, flood and invasion. And they allow us to glimpse lost worlds.

But this translation from pattern (fossil) into process (the organism and its biology) also requires the skills of a poker player. Imagine that we have sat down with three other explorers, say at the Luxor Hotel, where Howard Carter once dined, to play a game of cards, for access to our hidden treasure, the early fossil record. We know that we will need to win the game to survive. But the other players are not only poker-faced, they are downright mute. Worse—we are not told what game we are playing! Consider, for example, being given a hand of seven cards, ranging from Ace to King. Except that we don't know whether the Ace card is high or low, or whether spades trump hearts. Or whether there is a joker or not.

As soon as we are told the name of the game, or its rules, then all becomes clear and the game can be won. But we are never told what the rules are with the fossil record. We must therefore take risks—like a poker player—and begin by guessing at the rules. Then, step by step, we can hopefully iterate our game towards a fuller understanding of how to win. That is, to some extent, how decoding of a hidden message typically works. But remember this: explorers have only been sitting down at the green baize table of science, trying to decode the game of life, for about four hundred years or so. Yet four hundred years is a mere blink of an eye within the six million years of our existence as upright apes. Winning the game of decoding the early fossil record is also bound to be difficult in the first few rounds. As we shall see, it had to evolve from careful and prolonged watching for patterns on the one hand, towards inspirational hunches (that we call 'hypotheses') about processes on the other. Human progress towards learning the rules for decoding the fossil record has therefore been slow, requiring trial and error, with lots of questions, intuition and counter-intuition, accompanied by oceans of doubt. But then, science, which always rejoices in a good question, is a unique system for the measurement of doubt.

CHAPTER TWO

THE DEVIL'S TOENAIL

Ancient cups

I was catapulted into the quest to solve the dilemma of Darwin's Lost World in September 1973 by a single piece of rock. It was full of black and white swirls like 1970s pop art. This weird cobble had been discovered when my geological colleague Roland Goldring, jumped out of his Landrover to open a farm gate at Brachina Gorge in the outback of South Australia. That was back in 1967.[32] He had spotted it propping up a fence post and, without much thought, popped it into his tucker bag and brought it back to Reading in England.

By late 1973, I had gladly abandoned my highly unpromising career as a Survey Geologist and had pitched everything on a single throw—a temporary lectureship at Reading University, at that time one of the premier places in the world to study ancient sediments. I now needed a Big Problem to solve. Ideally, something as big as the dilemma that Darwin himself spelled out on page 306 of the *Origin of Species*: I fancied something to do with Darwin's Lost World.

Roland kindly put in front of me two really strange rocks. The first was a slab of rusty red sandstone from the Ediacara Hills with unusual disc-shaped impressions, thought to be the remains of Precambrian jellyfish. The other was the slab of rock from Brachina. Knowing that Martin Glaessner in Australia was reputed—incorrectly, as we shall see—to have decoded the riddle of the Ediacara markings, I literally sprang upon the Brachina tablet and carried it back to my room. For days on end, I held this curious slab in my hand, twisting and turning it, thinking and probing, trying to decode its hidden message.

I could see many visually appealing fossils within the slab, of the kind known as Archaeocyatha, which is Latin for 'ancient grail' (and grail, of course, is an ancient name for a wine glass). Archaeocyaths are really rather handsome fossils—in life their colonies must have somewhat resembled a squabble between Italian ice cream vendors—leaving cone-like shells scattered across the seafloor (see Figure 2) When they are split apart from the wine-coloured mud in which they are usually entombed, the bone-white shells of archaeocyaths can resemble fancy lace or flower heads like daisies—especially when the imagination is helped along with some excellent Australian wine.

Three things have for long absorbed palaeontologists about these 'ancient grail' cups. They first appeared near the base of the Cambrian, for no obvious reason. In other words archaeocyaths were, like trilobites, among the first conspicuous shelly fossils known—at that time—to appear in the whole of the fossil record. They then suffered a complete reversal in their fortunes, to vanish after a span of only a few million years, again for no clear reason. But best of all, nobody could agree for years exactly what kind of creatures they were made by. When I began work in this field, *les éminences grises* thought that they were primitive creatures allied either to single-celled amoebae or even to seaweeds. Some thought they were the extinct remains of multicellular animals,

Figure 2. The Tommotian commotion. Fossils from the earliest Cambrian rocks of Siberia, about 530 million years old, as reconstructed by the author. These include, in their rough order of appearance, and clockwise from three-o'clock, the cactus-shaped problematical fossil *Chancelloria*, the disc-shaped archaeocyath sponge *Okulitchicyathus*, the conical archaeocyath *Kotuyicyathus*, and the earliest trilobite *Fallotaspis* whose appearance marks the end of the Tommotian. These fossils are each typically less than 10 cm across.

perhaps belonging in their own Kingdom. And others thought they might be allied to the most 'primitive' kinds of animals that we have alive today—the sponges.

As we have seen within the lagoons of Barbuda, sponges consist of colonies of cells that are strangely casual in their connectedness, and they appear largely disinterested in body symmetry. The archaeocyath fossils from Brachina also displayed numerous small pores, much like the living sponges of Barbuda. But they had a strongly marked symmetry which is much more like that of living and fossil corals. Indeed, extinct archaeocyaths showed an obsessive, almost unhealthy, interest in precise geometry during their short dominion on the Earth, producing skeletons of breathtaking precision and elegance.

I decided to crack the riddle of these very ancient fossils for myself using a single large block of material from South Australia. I fancied myself, perhaps a tad presumptuously, as a young Champollion about to decode the mystery of Egyptian hieroglyphs. But Champollion only had a paper copy of the Rosetta Stone. I at least had a stone. And I was lucky in my choice. When I cut my rock into a dozen or more slices, it revealed two things seldom seen in the fossil record: archaeocyaths growing together in their positions of growth, and archaeocyaths competing with each other for space. Piece by piece I reassembled this 520 million-year-old community of creatures, and attempted to reconstruct how they had fought for space on the seafloor. The outcome of these ancient battles was preserved right there, in the rock. One ancient cup could be seen to settle on the sea floor, and then to start growing upwards around its rim. Then another ancient cup would glue itself onto the first one, and proceed to give it hell—the unfortunate host would grow deformed or even wither away. And then the invader would take over its living space.

Bones of contention

The ancient cups from Brachina provided what looked like the earliest fossil evidence for a 'fight' between competing organisms. It was a hair-raising thing to see across such a vast span of time, and revealed the rawness of animal behaviour, right from the beginning. But what kinds of animals were these ancient cups? To find out more, I needed to go in search of two experts—Françoise and Max Debrenne—in the Jardin des Plantes, Paris.

The formal gardens of the Jardin des Plantes are laid out rather like the board of a giant Monopoly game along the banks of the Seine. A row of enormous green glasshouses has been arranged along the west side of the gardens, to house the collections of tender and tropical plants. Facing them across avenues of trees and fountains are three rather grand buildings that make up Le Musée d'Histoire Naturelle—home to the collections of plants, minerals and fossils (see Plate 3).

The Fossil Hall is just one of these buildings. Its proper name is *Le Galerie d'Anatomie comparée et Paléontologie.*[33] This red stone tabernacle stands proud and alone, guarded along its outside by the sculpted stony gaze of a dozen dead naturalists. The hall itself resembles a very fine Victorian railway station. But it is a railway station filled to the brim with bleached white skeletons, all facing expectantly towards the ticket office. Every kind of mammal is displayed here, from giant baleen whales to tiny marmosets. And all of them, it seems, were cruelly stripped of flesh and frozen in mid stride while rushing to catch some long forgotten last train, back in the 1890s.

But the Fossil Hall is not just an assault on the senses. It provides a tweak to the intellect as well, if we know how to read it. The bones and fossils in this great building were put together by some of the greatest torch-bearers of the French enlightenment: Louis LeClerc de Buffon, Jean Baptise Lamarck, and Alcide d'Orbigny. But it is Georges Cuvier whose mind seeps out towards us from the bones here.[34] Cuvier was a pioneer of comparative anatomy, one of the great domains now absorbed into modern evolutionary thought. Collecting all the cadavers he could get hold of, and analysing them bit by bit, he found that he could read the way of life of each mammal from the shape and detail of its skeleton. A famous story, alas apocryphal, tells how one of his students, wishing to frighten old Cuvier by dressing up as the Devil, complete with goat horns and cloven hooves, ran into his bedroom crying 'Professor Cuvier, I am the Devil and I have come to eat you up!' Cuvier was, however, curiously struck by the benign anatomy of his tormentor: 'I am not afraid of you, *mon ami*', he replied, 'because you have both horns and cloven hooves. You must surely be a strict vegetarian!' So at least the story goes. It was this kind of logic that led scientists towards the realization that bats are not birds but flying mammals, and that whales are not fish but sea-going mammals. Each of these discoveries paved the way for the evolutionary thinking of Lamarck, and later for that of Charles Darwin.

Down in the dingy basement of the Fossil Hall in 1973, I was mindful of Cuvier's ghostly presence as I anxiously showed my archaeocyath fossils to Françoise Debrenne. Looking at this Brachina material, Françoise was also intrigued. Happily, she agreed that it helped to show that archaeocyaths had aggressive interactions just like living sponges and 'higher' animals, and quite unlike the seaweeds that some workers, such as Chicago palaeontologist Jack Sepkoski, thought them to be. But a second point was emerging, even more telling from the point of view of our story. As one of my students, Jon Antcliffe, has since put it: 'if you don't understand how an organism grows, you don't really understand anything about it.' Baron Cuvier was showing us, via his great hall of bones, that we can solve questions about the nature and relationships of enigmatic organisms such as archaeocyaths by comparing their patterns of growth and repair after damage. In other words, not only the presence of pores and canals but the sponge-like repair tissues in archaeocyaths indicated that they were animals, and that they were probably sponge-like animals.[35]

The Explorers Club

Ten years later, on a spring evening in May 1983, palaeontologists were starting to gather from around the world for a meeting in the grounds of a large country house, called 'Burwalls', outside the seaport of Bristol in south-west England. They had come together for a week to discuss, in private, the matter of Darwin's Lost World (see Figure 3). This meeting at Burwalls occurred during the very peak of the Cold War. The Soviet military, under their ailing leader Mr Andropov, had 11,000 nuclear warheads pointing towards targets in the West. The Americans had another 10,000 pointing right back. A movie about nuclear war, called 'The Day After', was being shown on limited release, making everyone jumpy. Much worse, President Ronald Reagan had been fanning the flames with a speech about 'The Evil Empire' and his 'Star Wars' initiative. The voice of British Prime Minister Margaret Thatcher provided a shrill *ostinato*, yet again. During the May of 1983, therefore, the powers of the East and the West were seemingly moving towards 'all out nuclear exchange'. Under these circumstances, it was fitting that our gathering of palaeontologists closely resembled the cast and stage set for a Cold War murder mystery: the

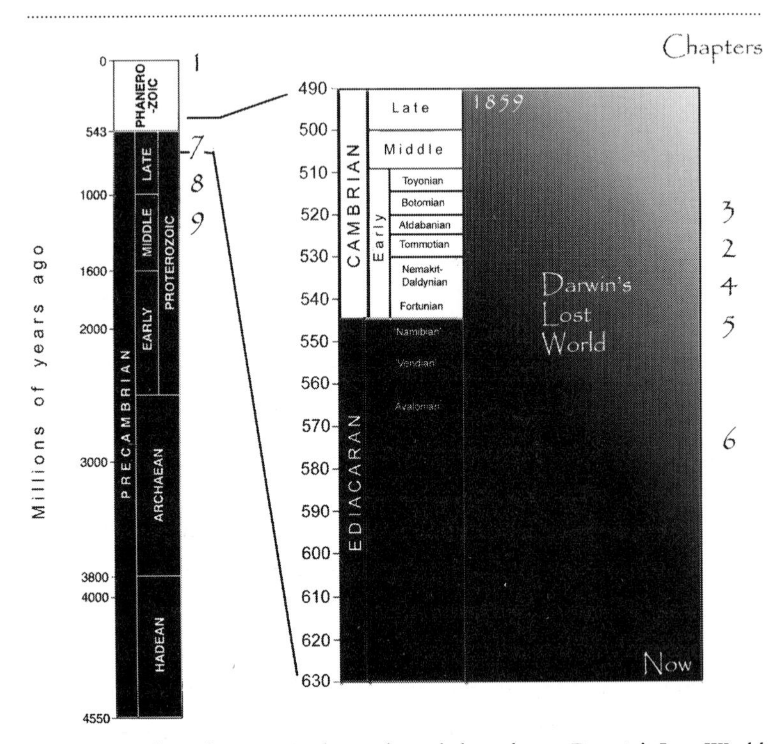

Figure 3. Telling the time. As this geological chart shows, Darwin's Lost World now encompasses more than 80 per cent of earth history and is here shown in grey and black. When Darwin published the *Origin of Species* in 1859, little was known of the fossil record of animals before the Late Cambrian, some 500 million years ago. The last fifty years has opened up the early fossil record back to nearly 3000 million years ago. Chapters 1–9 in this book move the story progressively back through time, roughly as shown by numbers 1 to 9 at the side of the chart.

'stage'—Burwalls—was an old Gothic mansion with pointy roofs and oak-panelled halls, erected by the Victorian inventor of 'Wills Whiffs Cigars', while the 'cast' was an oddball collection of eccentric professors, housemaids, Russians, Chinese, and 'translators'. Like victims in a murder mystery, the players were to be seen moving furtively, throughout the week, from one assignation to another: first in wainscoted rooms, next in the shrubbery; now in group-discussion, next in private huddles, communicating with each other in strange tongues. All they seemed to lack was an in-house detective.

But they—or rather we—did have a body. In fact the gathering had thousands of bodies: the fossil remains of the Cambrian period which either lay before us in trays, or were concealed in our heads, or were even more tastefully hidden from view in our publications. The aim of this gathering was nothing less than global agreement on a first narrative for Darwin's Lost World–called *the Precambrian–Cambrian Boundary Problem.*[36] A tall order indeed, given the strong personalities present, but founding father John Cowie was hoping for a breakthrough. Here, then, was a chance to come to a decision on what was understood about the difference between life in the 'Cambrian' and life in the 'Precambrian'. Not only that, but the aim was to find a place somewhere in the world where this dramatic transformation in the history of life could best be examined. There had been many earlier attempts to get to the bottom of this, including one meeting in Prague of 1968, hastily dissolved after the arrival of Russian tanks in the city. The Russian tanks were not there to claim the Precambrian–Cambrian boundary, of course, though we could be forgiven for thinking so because, as we shall see, this boundary is arguably the greatest division in the whole of Earth's long—4,560 million year long—history. It was, and still remains, a great prize. It divides the familiar world of the Cambrian from the long dark night of the Precambrian (see Figure 3).

For most of the century of research that was to follow Darwin's publication of the *Origin of Species* in 1859, it was widely believed that the great Precambrian 'dark age' largely came to an end when those long-extinct creatures called trilobites first appeared on the stage. So common are trilobites in many rocks of the Cambrian that their evolution can be used to correlate Cambrian events—such as climate and sea-level change—across the planet. The Cambrian period had therefore been widely regarded as the Age of Trilobites for generations. And one of the best places to study early trilobite evolution lies high up on the vertiginous slopes of the Mackenzie Mountains of the Yukon and North west Territories in Canada.

In the Bristol lecture hall, geologist Dr Bill Fritz explained to the gathering how he had spent the previous decade mapping the rocks in this remote wilderness, braving blizzards, grizzly bears, and helicopter dramas. He had then helped to trace the evolutionary history of the earliest trilobites along the great spine of the Rocky Mountains of North America from the Yukon down into the White Inyo Mountains of California, where they nestle among the Bristle Cone Pines. What was this world of trilobites actually like? To find out, let us take a closer look at this first American trilobite, called *Fallotaspis*.

First shield of defence

Fallotaspis is not very large as trilobites go,[37] about one or two centimetres in length. But it is rather striking in appearance (see Figure 2). At one end lies a half-moon shaped head shield with a prominent pair of crescent-shaped eyes. And behind the head streams a long spiky body divided into many segments that helped it to crawl and burrow. To some, it resembles a spiky terror from a horror movie. I have even seen stout lads recoil in horror from such fossils: our human fear of arthropods is prehistoric, and hard-wired into our brains. This arthropod fossil called *Fallotaspis* can be traced from America across the Atlantic into the Atlas Mountains of Morocco, as well as along the great river systems of Siberia. That is rather curious because there is no evidence that *Fallotaspis*—which means the shield of Monsieur Fallot in Latin—was ever a strong swimmer. It is not found in Cambrian deep-water sediments in those same areas. This has been taken to suggest that North America and Morocco, which are now so far apart, must have been much closer together in the Cambrian. Indeed, such evidence reveals that the modern Atlantic Ocean did not exist at all at that time.

Trilobites were highly successful in their day. The earliest forms, such as *Fallotaspis* arguably owed their winning ways to the presence of rather sophisticated eyes. Indeed, trilobites were arguably the earliest creatures to have looked out upon their world. Light was captured by means of a pair of semicircular eyes placed near the middle of the head shield. When well preserved, these great crescent-shaped eyes can display tiny, honeycomb-shaped lenses, all packed together closely like those of a modern house fly, and they may well have glistened on the seafloor. These eyes were not, it seems, needed to look out for food. Instead, they appear to have looked upwards and outwards, perhaps nervously surveying the Cambrian ocean for any predators that might dart in from above.

On the underside of the head shield of *Fallotaspis* was found a simple mouth into which worms and sediment could be shovelled with the help of seventeen pairs of long spiny thoracic limbs. Although these limbs are seldom preserved in the fossil record, they appear to have been anchored in rows strung along the body. To any predator attacking from above, the underside of *Fallotaspis* would have presented a delectable but invisible morsel, like the underside of a shrimp or a crab today. Such a trilobite would have moved about the seafloor in much the same jerky way as a shrimp, trying to keep its soft underbelly out of sight, and taking special care not to become turned upside down.

The tasty, 34-legged *Fallotaspis* had also evolved two further means of defence, to stop it appearing too frequently on the menu. Both were highly innovative. First, the animal had developed a hard mineral crust of calcite (like a lobster shell) on its top surface. That meant that any predator would have to be strong enough to turn it over if it ever wanted to dine upon the soft and tasty underbelly. Second, when even this tough shell-suit failed to deter, it could burrow rapidly into sediment with its powerful limbs, keeping the underlying soft parts such as gills out of harm's way.

Both of these clever strategies left their mark upon the early Cambrian fossil record, about 525 million years ago. The hard shells are, of course, the fossils found in the rocks. And the burrows—called *Rusophycus*—can be found as scratch marks preserved on the bottom layers of sandstones wherever *Fallotaspis* and its relatives lived and played.

But is it true that trilobites were really the first animals to emerge from the Precambrian 'dark age'? To find out, it is time to return to the lecture hall at the Burwalls meeting.

Tommotian commotion

This meeting in 1983 was to become legendary, as we shall see, for its Cold War manoeuvring. Teams from Moscow, Siberia, China, North America, Australia, and Europe were jockeying for pole position. Their aim was to review regions of the planet that were often so remote or forbidding that few, save the speakers and their teams, had ever been there. These regions were to include the vertiginous mountain slopes of the Yukon Territory in Canada where we have just been; the White Sea coast of Arctic Russia; the upper Yangtze Gorges of China; and the fog-girt peninsulas of Newfoundland. There was a hidden pressure too, upon the Communist delegates, to return home with the prize—the Precambrian–Cambrian boundary. That pressure was to keep the organizers (including myself) on the very tips of our geopolitical toes.

Hitherto, teams from the Soviet Union had dominated thinking about Darwin's Lost World and the Cambrian Explosion. That is partly because both Cambrian and Precambrian rocks are unusually well preserved in Russia. Darwin had already drawn attention to this back in 1859, saying: 'the descriptions which we now possess of the Silurian [Cambrian to Silurian] deposits over immense territories in Russia and North America, do not support the view, that the older a formation is, the more it has suffered the extremity of denudation and metamorphism.'[38] Curiously, this prominence of Russian rocks was to become greatly amplified as a consequence of the Bolshevik Revolution back in 1917. That may seem surprising to us now, but one of the engines driving forward geological exploration during the 1960s to 1980s was the battle between capitalist and communist systems and their respective mineral and hydrocarbon exploration programmes. Because of the urgent need for countries to develop resources in the battle, geologists in both Russia and China often received high status. Whole cities of scientists, such as Novosibirsk in Siberia, were set up for the exploration, development, and colonization of the Arctic lands, especially in the north-east. Geologists would map far and wide during the summer, deep into the lands of Siberia. Palaeontologists would labour throughout the year on the materials brought back to their laboratories, naming more and more new fossils as they were discovered beneath their microscopes, set up in Novosbirsk or Moscow.

Huge advances were made in this way by Soviet scientists into the nature of the Precambrian–Cambrian transition. At the Bristol meeting, Dr Alexei Rozanov proceeded to tell us, in his pleasingly rolling Russian accent, about decades spent mapping and exploring in Siberia.[39] Russian palaeontologists had made one very important discovery, as well. They had stumbled upon a period of time before the appearance of *Fallotaspis*, in which other kinds of skeletal fossils could be found. It was his team that conjured up the term Tommotian, to describe an early world dominated by 'small shelly fossils' that had lasted from about 530 to 525 million years ago.

It is important to emphasize that no trilobites have ever been found in rocks as old as the Tommotian. They seem to have been either unpreserved or absent. But many other animal groups had made their debut onto the stage of Earth history by that time, including a few still found living along the seashore today, some 530 million years later. Tommotian 'small shelly fossils' typically include worms, clams, snails, and sponges. In other words, the Tommotian world was stocked with many organisms rather like those we have just met in the modern lagoons of Barbuda. To take a closer look, I therefore decided to travel to Siberia.

Inside the Labyrinth

In the summer of 1990, I found myself edging slowly towards the fringe of a Russian forest, to be motioned to halt by an armed guard, holding back an Alsatian dog on a rope. Nodding sternly, the soldier locked the great iron gate behind us and indicated that he had instructions to escort me through the woodland that surrounds and conceals the gaunt buildings of PIN, the Moscow Paleontological Institute. The soldier, the dog, and I walked briskly for a mile or more beneath a gloomy canopy of sycamore and ash that sheltered us from the rising warmth of the Moscow morning. I had come in search of witnesses, and clues, that could lead us towards the truth about Darwin's Lost World. Rozanov and his colleagues had shown that the answers to our questions lay hidden in rocks far away, along the icy river banks of Siberia. But to get there, I had first to grapple with dragons.

We entered the famous great hall of dinosaurs at PIN.[40] This cavernous room had been closed to the public for months. Not only that, but all the lights were switched off and the great bones were draped in white sheets. Beneath these shrouds lurked the remains of some the finest dinosaur skeletons ever excavated in the Gobi Desert of Mongolia. These white sheets, I was soon to learn, were a sign of the growing breakdown of economic conditions in Moscow. An acute shortage of funds had caused PIN to sink the halls into darkness, and to hide the monstrous bones beneath dust sheets. Or even, allegedly, to spirit them away for funds. These were indeed dark times in Russia.

The drapes made the dinosaur skeletons beneath look menacing, an effect increased by the iron hinges of the great wooden doors around the atrium—one resembling the horns of a giant elk, another like the pincers of a scorpion about to pounce. But such fantastical musings were soon displaced by more prosaic doubts about the wisdom of travelling into Siberia at such a time of trouble. I passed through the 'Jurassic' hall and climbed the steps up to the famous labyrinth of dimly lit rooms and dusty corridors of PIN. The upper corridor was narrow, and lined with pine wood shelves that bulged with fossils and bones from floor to ceiling. In the drawers were concealed some very old fossils indeed, much older than those of the Gobi dinosaurs—up to half a billion years older. The labels on the drawers, in Cyrillic script, showed that they had been collected from Siberia, from among the colonies and old prison camps. They had been hacked out of rocks that spanned the vast gloom of the Precambrian.

My mind continued to race. Would Director Rozanov, allow me to see some of the key fossils that might lead to the rediscovery of Darwin's Lost World? Rumour certainly had it that the Russian rocks were yielding animal fossils that were both very old and very strange. But when fossils are both old and strange, their Keepers will tend to keep them hidden under wraps. That is not surprising, of course. Names and reputations can be made from the discovery of an unexpected ancestor—such as the first fossil fish—if the tale is cleverly spun. But another worry lurked, since it was rumoured also that conditions in Siberia had been so bad that year that parties of geologists had come back to Moscow emaciated and broken from an epidemic of *Giardia*, a particularly miserable form of dysentery. Nor was I the only geologist hoping to get to Siberia that year. Others were converging from all round the world, hoping to take part in this rarely offered opportunity. And we began to hear that petrol was getting very scarce. It was certain that some of us would be prevented from getting to the rocks that held the clues. I could only hope it would not be me.

That was still during the Cold War, so nothing was straightforward. Access to the rocks might be denied for other, more geopolitical, reasons, too. Over preceding years, Rozanov and his team in Moscow had poured huge amounts of time and energy into exploration in Siberia. They had constructed an elegant case for a major radiation of the animal groups at the base of the Tommotian Stage along the Aldan River. But a second group, largely confined to Novosibirsk in Siberia, was arguing for a rather different story. Dr V. V. Missarzhevsky and his colleagues claimed to have found rock sections that lie much further north (along the Anabar River) or further south (in remote 'Outer Mongolia'). They argued that there the animal skeletons had been found at even lower levels, in a distinct period of time, called by some the Nemakit-Daldynian. Emotions were running high on this issue, not least because the latter group felt that their data was not getting a fair hearing—they thought they were being over-ruled by Moscow. Director Rozanov countered that the Nemakit-Daldynian could not really be distinguished from the Tommotian—it was simply that the rocks looked different from one place to another.

This really mattered then, and it still matters now. Was the Cambrian explosion of animal life very abrupt, starting mainly at the base of the Tommotian, or did it perhaps begin many millions of years earlier, near the base of the Nemakit-Daldynian? With colleagues around the world, I had been developing a method to attempt to solve this, a way by which we could hope to test these two competing hypotheses. It would involve measuring and comparing the carbon chemistry of the rocks themselves. But it would also require negotiating a veritable maze of bureaucracy. I would need collaboration from the various groups. And, most of all, I needed cooperation from the Director himself. As it happened, the Director plied me with mugs of tea, and I was greeted with smiling faces around his office.

But just when all seemed to be going well, the worst happened. My visa to Siberia was seen to have a serious error on it. Instead of stamping 'Yakutsk', the customs office had stamped 'Irkutsk', some thousands of miles to the south. Hopes of getting to see Darwin's Lost World were suddenly in jeopardy. It was far too late to do anything. So I walked back through the forest to suffer a sweaty, sleepless night, angrily slapping Moscovian mosquitoes on to the walls of my room with a wet bath towel.

Ulakhan-Sulugur

Over a week later, a pitiful cluster of scientists was sitting huddled together in the cramped cabin of a Soviet launch, its decks draped with ice. The roar of the engine as we steamed northwards along the river Aldan, was muffled by the buffeting of waves, the shipping of spray, and the clatter of hailstones from a passing icy squall. Our aim was to get to a little beach along the shore called Ulakhan-Sulugur. We held onto our baggage and stared glumly out of the steamy portholes at the passing scenery. For mile after mile, we pitched and tossed our way along the channel. The river Aldan seemingly wound on forever between dark and forbidding glades of Siberian pine. No grass. No flowers. Just stands of timber that pressed right down to the river's edge.

It had taken a week or so of negotiating and travelling to get here from Moscow, on an old propeller plane, a helicopter, and boats like this. Our first stop was in the old prison town of Yakutsk. Here, we found that pavements were bulging and cracked with ancient tree stumps that were working their way upwards from the permafrost beneath. Concrete accommodation blocks were connected by huge overland pipes that, in the long winter, carried the heated steam needed to keep life going. Our first sleep was taken in an abandoned school in the forest. Thence we were transferred to our very own concrete block where a canteen, serving a dish of 'cadaver and rice', made even my 1950s English school dinners seem edible. Some days later, we were herded into a huge orange Aeroflot helicopter, that flew for much of the day in long hops towards the Aldan river far to the east. The helicopter had no doors. For that matter it had no seat belts, and no real seats. But it did have benches that were tilted menacingly towards the parachute drop in the middle—that would never have been allowed by the Oxford University Health and Safety at Work Committee. And the pilot seemed to have his feet on the control panel and a bottle of vodka in his hand. I had never flown in a helicopter before, and innocently thought this was normal. But my chopper-wise Canadian friend, Guy Narbonne, remained unusually pallid throughout the flight. Apparently, this was not normal.

At last we saw a huge cobble island in the middle of the Aldan River. The helicopter circled to drop us down in the middle of absolutely nowhere; a nameless place of stones. We tumbled gratefully out of the helicopter with packs on our backs. The pilot then shouted something in Russian. The huge black wheels of the chopper began to glide sideways, grazing my prostrate body and pressing it hard to the cobbles. I had never flown in a helicopter before, and again thought this was normal. Apparently, this was not normal either. The pilot was shouting for us to 'duck!'

The sound of a helicopter rumbling away into the distance is one of the most haunting sounds that a modern explorer can hear. It is balanced only by the joy of hearing its rumble return again, some weeks later. For a minute, we tried get our heads around the fact that we were marooned in the middle of an island, in the middle of a river, in the middle of Siberia. Then we espied a launch coming to pick us up for our transfer to the campsite some miles upstream. By this time, we were dehydrated from a diet of 'dogfood' and vodka, tumultuous travel, and a lack of sleep. We had dossed down in old prison towns, empty schools, and forest floors. We were sick from stomach cramps. And we were cold and damp. There we were, thousands of miles upstream from the point where waters eventually pour into the Arctic Ocean to the north, and almost as many miles from the nearest civilization to the south. Only weeks before, I had been sitting in the comfort of my college rooms in Oxford, sipping sherry and reading *Country Life*. The following morning, it was still not clear whether I would be allowed to join in with the long dreamed-of expedition up the Aldan. Of the sixty or so geologists only a handful would be permitted to make the journey.

Next morning, however, the lobbying in Moscow seemingly began to bear fruit, and the committee selected me to take part. So there we were, a dozen of us from around the world—including Alexei Rozanov from Russia, Xing Yusheng from China, Gonzalo Vidal from Sweden, and Guy Narbonne from Canada—pounding along in the launch beneath leaden skies. Our discomforts were soothed now and then by the passing of vast great cliffs of layered dolomite. At times these cliffs gleamed white in the Arctic glare, then turned to dun-coloured in the gloom. Each of these cliffs, some of which rise up to fifty metres in height, looked like a vast stone wall, eaten away by eons of meandering and erosion. And it was these great cliffs that we had come to read.

But not any cliff would do. With a practised eye, we scanned the beds of rock to confirm that they were gently dipping northwards. This meant that higher and younger layers of ancient sediments laid bare in these cliffs were progressively being revealed before us—like the unfolding story in the pages of a great book. We had passed by half a dozen or so white cliffs along the river during the last three hours or so. And only then did our eyes behold a more distinctive cliff line as it rose out of the water towards us. This bluff of rock lay along the left bank of the Aldan. It seemed curiously different from the others. The lower beds of familiar white dolomite were capped by beds of startlingly red limestone. My heart leapt! Here, at last, was the fabled cliff, our destination—Ulakhan-Sulugur.

We jumped down onto the beach. I was tempted to kiss the ground. As it happened, I was obliged to do just that—in a dysenteric sort of way. Waves of nausea and vomiting left me weak and trembling on the shoreline.[41] In the background, the engines of the launch were switched off to save diesel fuel (see Plate 4). The captain was anxiously debating the fuel reserves with his crew—after all, this was 1990, the year of perestroika and the collapse of the Berlin Wall. As our guide Alexei Rozanov would say: 'anything possible—and everything impossible—in Russia today'. It was then that our ears were assaulted with the silence of the Siberian forest. We could hear little more than the lapping and lilting of waves against the hull of the boat and the rustle of icy wind in the treetops above the cliff. No birdsong. No crickets. Just an unsettling, brooding silence.

After a respite, we were clambering on to the rocks at the foot of the great cliff, bashing away with hammer and chisel as though our lives depended on it. And in a way they did—at least our professional lives. This was a brief, elegiac moment of harmony, within a career spanning four decades. Only months before, I had been reeling from an acrimonious exchange with a large and usually jovial Swedish geologist called Gonzalo Vidal. But here, in fossil Mecca, our differences dissolved and we shared our toils together at the rock face. A minute's worth of bashing brought down slabs of greenish grey rock covered in fossils—brachiopods, hyoliths, and the first archaeocyaths. Suddenly, we were like fossil hunters entering fossil heaven. These 'small shelly fossils' were some of the earliest evidence for the convincing presence of animals anywhere in the world. On the previous day, we had hammered away at the underlying, older dolomite rocks upstream but had found next to nothing. But at this higher point in the rock succession, everything seemed to change. The fossil record, after literally billions of years cloaked in secrecy, was suddenly having its own perestroika moment—one of the greatest revolutions in the history of life: the 'Cambrian explosion'. And let me emphasize again that the Cambrian explosion was not normal. It was decidedly odd. Almost the oddest thing that has ever happened in the history of our planet.

The Devil's toenail

The transformation from the Precambrian to the Cambrian world, so vividly displayed in the cliffs at Ulakhan-Sulugur, is the biggest dividing line in the whole of Earth history—much bigger than that between man and the other apes and almost as big as that between life and pre-life itself. It is hard to overstate its significance. And of all the fossils that fall out of the rock at Ulakhan-Sulugur, none better exemplifies the oddities of the Cambrian explosion than does a little fossil called *Aldanotreta*.

It looks like an old toenail accidentally preserved on the seafloor. It is not very attractive to look at—like a Devil's toenail perhaps. But descendants of these fossilized toenails are still living, and minding their own business, on seafloors around the world today. Ulakhan-Sulugur marks their first entrance on to the grand stage of life. Even stranger, many of them have since remained virtually unchanged for nearly 530 million years! They came, they saw, and they calcified. It also seems likely that they will still be here on Earth long after we have been and gone. These charming little 'toenail beasts' are called brachiopods.

Brachiopods do not, at first, seem to be very exciting organisms. Even after a second look they can seem just a little disappointing. I used to tease my students during lectures with a still colour slide of a living brachiopod, and then kid them along that they were watching a video. They would watch the screen enrapt for a minute or so, and then become restless when nothing happened. Maybe that was a bit unkind of me. But that little vignette of 'a day in the life of a brachiopod' was intended to enshrine a curious truth: nothing much happens. And the bigger picture of brachiopod history is also a bit like that as well. Nothing much happens. Paradoxically, that is a measure of their very great success. Only 'winners' get to stay forever in the charts, somewhere. Brachiopods are like the Cliff Richards of evolution.[42]

So why is our little brachiopod *Aldanotreta* so iconic? For two curious reasons. First, there is the presence of a hard outer shell—the toenail bit. This shell enclosed a space that enabled the little brachiopod inside to feed. It is almost impossible to imagine any brachiopod without its shell. Like a pop star without a Cadillac, it would be missing the essential ingredient that helps it get to work. This is interesting because *Aldanotreta* was one of the first organisms to acquire a shell, and possibly the first brachiopod to do so. The Cambrian explosion is pre-eminently about the acquisition of shells and other kinds of skeleton.

The second curious reason for *Aldanotreta*'s celebrity is that it is one of the oldest fossils that we can find whose remains can be attributed to an animal group that is actually still alive today.[43] Its closest living relative is a little oval shell called *Lingula*, still found living around the shorelines of Japan. Many fossils older than *Lingula* and *Aldanotreta* are known from the rock record, as we shall see. But most are not easy to relate to any living group. In other words, the modern living world had its debut in the Tommotian, within the cliffs of Ulakhan-Sulugur. The next 530 million years of life were to see their little ups and downs. But the same animal groups have stayed with us throughout that huge span of time. And by 'us' I here mean all the animals with backbones—from fish to ourselves—that belong to the so-called 'chordate phylum'.

The cup of Okulitch

In the red limestones that lie above the level with the 'Devil's toenails', there were many other strange shapes embedded within the cliffs. To find better examples, we had to travel back to the base camp at the foot of the great cliffs of Dvortsy. Some of these fossils looked like trampled wine glasses, seemingly left behind by the orgy of life that began in the Cambrian. Even the rocks themselves were stained wine red. Climbing up the cliffs for a closer look, I could see that these goblets were the remains of our old friends, the archaeocyaths. Here, however, these sponge-grade fossils abound in rocks that are some five million years older than the first trilobites like *Fallotaspis*.

Archaeocyaths first turn up at the very base of the Tommotian in Siberia, almost exactly at the point where the rocks turn from buff-coloured dolomite to wine-red limestone. These Tommotian forms have the simplest shells as well, with simple pores like the kind called *Okulitchicyathus*, after a great Russian palaeontologist (see Figure 2). A bit of scrabbling about on the rocks shows something else rather remarkable too. The ancient ecology here must have somewhat resembled the one we have just seen in modern Barbuda. For example, modern Codrington lagoon has long ridges of mud that are colonized and stabilized by seagrass and algae, on the tops of which live sponges and molluscs. Almost exactly the same thing can be seen here in the rocks at Dvortsy. The archaeocyath skeletons are clustered around greyish mud mounds about two metres high and three metres across, which geologists call 'bioherms'. In other words, the Tommotian seafloor was rather like that of Codrington Lagoon—limey muds in waters that were not very deep, with a seafloor covered in masses of shells and algal remains.

But there was another surprise in store: the rocks that underlie these Tommotian spongy mud mounds consisted of buff Yudoma dolomites with plentiful evidence for microbial laminations but very few animal shells, just like the waters of modern Cuffy Creek. With a little bit of care, it was possible to reconstruct the environmental story here. The lower rocks had features that looked like the very shallow salt ponds and microbial mats of Cuffy Creek. And the higher wine-red rocks had the characteristics of deeper water marine lagoons and embayments, with algal mounds, like modern Goat Reef. If that were so, it could be argued that the sea must have advanced over the lagoons and land surfaces at the start of the Tommotian.

And if the sea had advanced over the land in the Tommotian, it could mean that the explosion of animal fossils in this section of Siberia may not have been a real evolutionary explosion at all—it could have been due to environmental change. For example, hostile conditions of saline creeks with microbial mats and very few animal remains—in the Yudoma dolomite—seemingly gave way, in higher beds, to normal marine conditions with many fossils—in the Tommotian limestones.

Was this environmental reconstruction a true one? If so, then it would point towards an even stronger prediction: that animal skeletons older than the Tommotian could be expected to turn up in other ancient habitats, where the conditions were less hostile. Conversely, if the explosion was almost instantaneous—then there should be no evidence for earlier remains. Was there anywhere in these Siberian cliffs where we could test that?

An odd-ball

Happily, Gonzalo Vidal, Guy Narbonne, and myself found out that there was indeed a spot in the cliffs at Dvortsy where pre-Tommotian fossils could supposedly be found. Needless to say, it was in a spot that seemed rather inaccessible.

After breakfast the next morning, we therefore assailed the great slopes with the hope of reaching a small pocket of fossils that lay within rocks that were lower, and therefore older, than the Tommotian. The upper parts of the steep cliffs were covered with dense pine forest, and the view from the tops of these cliffs, once attained, was indeed magnificent. But our gaze was quickly distracted by our guide, pointing to a crag beneath our feet. A huge bluff of dolomite could be seen jutting out from the slopes here. Cautiously, we edged our way down the carpet of loose rocks to this bluff, until we saw a greenish layer that supposedly lay within the Yudoma dolomite. Here, we were assured, could be found the first signs of animal life in this section—thorn-like rays of the cactus-like fossil *Chancelloria* (see Figure 2). The setting looked worryingly like a local swallow hole on the ancient seafloor. But we filled our sample bags full of the greensand, ready for processing in the laboratory back home.

Chancelloria is an odd-ball fossil known only from the Cambrian period. Well-preserved examples have been found in the Burgess Shale of British Columbia as well as in rocks of similar age from Utah, where its cactus-like body can be seen to be covered with star-shaped rays, rather as though the spines of a prickly pear have been left behind after the cactus itself has decayed away to nothing. At Dvorsty, however, we were only looking for broken pieces of this shelly armour, in the form of a so-called 'small shelly fossil'.[44]

The thorn-like rays of *Chancelloria* can look, at first, like the sponge spicules found on the floor of Codrington Lagoon. But calcareous sponge spicules are typically solid, while the rays of *Chancelloria* are hollow, each provided with a little pore at the base. It was as though the creature had been trying to insert little digits inside the rays, like the sticky fingers of a child inside a glove. That doesn't sound at all like a sponge. Indeed, it doesn't sound much like anything else alive today. It resembles, as we shall see, attempts made by the earliest molluscs—simple limpet-like creatures such as *Maikhanella*—to secrete shells made from clusters of hollow spines. But *Chancelloria* was not at all limpet-like. Instead, it appears to have stayed rooted to the seafloor, growing upwards like a cactus, and adding new rays in circlets near its top.

Some fossils at the base of the Cambrian we can decode from comparative studies with living animals, like *Aldanotreta* and its brachiopod relatives. And some fossils seem only a leap away from living examples, like the fossil archaeocyaths and their living sponge relatives. But there are some fossil organisms in these ancient rocks which we can't really claim to understand at all. *Chancelloria* is such a fossil. These strange cliffs along the Aldan River therefore divide a younger world, which we can immediately recognize, including many familiar animals alive today, from an older darker world—before the Tommotian—in which nothing much seems to make sense. Not at first glance, at least.

The first trick

So what are those clues in terms of *pattern* that we have so far seen tumble out of the cliffs along the river Aldan in Siberia? Charles Darwin's first complaint, remember, was that fossil animals appear very abruptly—perhaps too abruptly—in the rocks of the Cambrian then known to him. As we have just seen, however, in the more recently discovered sections of eastern Siberia, the Cambrian explosion did not happen 'all at once'. Like the arrivals at Cinderella's Ball, the players came in dribs and drabs. First to arrive at The Cambrian Ball were some eager but ugly sisters, like *Chancelloria*. These were followed by more familiar animals like *Aldanotreta*. And last to arrive at the Ball, like Cinderella herself, were charming young trilobites with big brown eyes, such as *Fallotaspis*.

Our second big clue from Siberia was this one: that phases within the Cambrian explosion did not happen everywhere at the same time. There were strong environmental controls upon the appearance of our animal ancestors. In places where the physical conditions were tough, such as deltas, creeks, and mudflats, the appearance of animals was 'delayed' until conditions got somewhat better. That being so, this explosion probably began in warm, shallow seas and was unlikely to have been due to a sudden invasion from a hidden lake or lagoon.

To find out what was going on before the Tommotian, we therefore need to travel far away from these ancient creeks and mudflats of Siberia, towards a place where the early seafloor conditions could have been more benign, and where clues to the end of Darwin's Lost World might therefore still be found in a state that is rather more complete.

CHAPTER THREE

A FOSSILIZED JELLY BABY

Chengjiang

A group of us was standing by a hole in the ground, chatting innocently about trilobites. Suddenly, we were startled by a whoop of joy: 'Wow! Come and look at this!' Palaeontologist Vibhuti Rai had been hammering away at the honey-coloured mudstones when out leapt a fossil in a billion—it looked like a shrimp, with all its soft parts preserved. We crowded round to take a closer look. Coated in rust-red iron oxide, it was replete with antennae, gills, legs, and a tail. It even had little eyes on stalks. My own eyes grew on stalks in response. For a brief moment, we viewed Vibhuti with admiration. Then with something close to envy. And then our mood turned ugly. We began bashing away at the cliff like demented dwarves, each fighting for space along the ledge.

A few hours earlier, we had been sitting amicably enough in a wagon, bouncing along a pot-holed road to the east of Kunming, in south China. It was September—the time of the rice harvest. Few other vehicles were on the road that day, so the black tarmac was strewn with carpets of brown rice, set out to dry in the still warm sun. Water buffalo were in harness for threshing here and there but, by 1992, were already being replaced by chuffing tractors. Mile after mile along this road, our bus wound upwards through hills of red and brown mud towards the village of Maotianshan. Here, in a quarry several years before, geologist Xian-guang Hou of Yunnan University had made a stupendous discovery—Cambrian fossils with all their limbs and organs preserved.

Some comparison might be made between Hou's creatures from Chengjiang[45] and those of the Burgess Shale, made famous by Charles Walcott in the early 1900s and again by Stephen Jay Gould in the 1980s.[46] These Chengjiang fossils were also respectably old—early Cambrian—much older and arguably even sexier than the Burgess Shale biota. Indeed, they were only slightly younger than the rocks we have just seen at Ulakhan-Sulugur in Siberia. Graciously, we had been given official permission to collect fossils that afternoon.

The quarries at Maotianshan resembled a chain of bomb craters blasted out of yellow rock. Each crater was surrounded by a mound of shaley debris. Not very picturesque, perhaps. But when small blocks of creamy smooth mudstone were tapped firmly with a hammer, they yielded up amazing treasure. Fossils began to emerge, each shining out from the muddy matrix like the heraldic blazon of a flag—*Cindarella, gules, passant guardant within a field Or*—blood-red fossils set upon a field of gold.

Cindarella is a fascinating example of a Chengjiang fossil, and not just because of its fairy-tale name. Shield shaped, and two centimetres long, it looks like a parboiled, big-headed prawn. Unlike the trilobite *Fallotaspis*, or its relative *Eoredlichia*, this ancestor lacked calcite armour-plating. And like its ugly sister *Xandarella*, it shows what true trilobites seldom show—their rows of walking legs and feather-like gills strung out in two rows along the well-segmented belly. These little legs and gills are the giblets we would normally remove when shelling prawns in a restaurant. It is hard to avoid wondering, therefore, whether *Cindarella* tasted good when marinaded in Tabasco sauce and served with a bottle of white wine. Alas, we shall never know.

Another fine shard of shale revealed an old friend—a brachiopod called *Lingulella*. Just like *Aldanotreta* from Siberia, this fossil also has a toenail-like shell made from calcium phosphate. But it differs in preserving the remains of a fleshy attachment stalk, called a pedicle, a feature hardly ever preserved in other fossil brachiopods. Interestingly, *Lingulella* from the Cambrian of Wales can be blamed for greatly vexing Charles Darwin back in 1859, when he said 'Some of the most ancient Silurian [Cambrian] animals, as... *Lingula* &c., do not differ much from living species,'[47] and again, 'Species of the genus *Lingula*, for instance, must have continuously existed by an unbroken succession of generations, from the Lowest Silurian [Cambrian] stratum to the present day.'[48]

The puzzle for Darwin was then set out by him as follows:

> and it cannot on my theory [of evolution by natural selection] be supposed, that these old species were the progenitors of all the species of the orders to which they belong [here meaning the lingulide brachiopods], for they do not present characters in any degree intermediate between them.

By this, Darwin was saying that he thought *Lingulella*—as it is now called—was not only a living fossil. It was much too advanced to be the Mother of All Brachiopods. He was implying that the ancestors of complex animals like *Lingulella* and, of course, the trilobites, must have extended far back into the rock record, perhaps as far back as 1200 million years ago.

Charles Darwin was also extremely sceptical about the chances of finding fossil fish below the Silurian:

> Seeing...that the oldest known fish...belong to their own proper class...it would be vain to look for animals having the common embryological character of the Vertebrata, until beds far beneath the lowest Silurian strata are discovered—a discovery of which the chance is very small[49]

Imagine his reaction, then, had he known that the Chengjiang biota was also to reveal what some believe to be the earliest fossil fish. That would, perhaps, have led him to increase his expectation of a long and slow-burning fuse to the Cambrian explosion. These fishy fossils have lately emerged from a long sleep in rocks further north, near to Haikou. Here, the mudstones have received much less attention, perhaps because they are not bright red but dull grey in colour. In fact, this means they are actually better preserved, with films of the original organic matter in place. The bedding planes at Haikou are literally covered in hundreds of elongate greyish markings called *Haikouella*. This fossil looks very much like a modern jawless fish, such as *Amphioxus*. With a cigar-shaped body some two to three centimetres long, it had a pointed head provided with seven ovate gill slits, plus a dorsal fin and a long tail for propulsion. *Haikouella* even seems to have lived in shoals of several hundred, like sardines. *Haikouella*—and all the other creatures of the Chengjiang biota—then seem to have suffered in multiple events of mass mortality.[50] Their death is thought to have resulted from storms that stirred up poisonous hydrogen sulphide on the seafloor beneath. It is sad to think that this prettiest of Cambrian pictures was so cruelly painted.

Worms in drag

One of the most curious features about the Chengjiang biota, is not the oddness of what is here but the oddness of what is not. I am referring to the seeming absence of annelid worms—those ancestors of garden earthworms and seaside lugworms. We tend to think of these animals as almost ubiquitous today. And they turn up in goodly numbers in the younger Burgess Shale of British Columbia, in the form of fossils like *Canadia* and *Burgessochaeta*. It is hard not to smile at the sight of these old Canadian fossils, which can seem a mite overdressed. *Canadia* must have looked like a worm in drag—all ruffs and frills, flounces and furbelows (see Figure 4).

No signs of any old worms in drag, though, in the muds of Chengjiang. Instead, the commonest fossil here resembles—how can I put this nicely—a novelty condom complete with its organ intact. This fossil, called *Paraselkirkia*, was appropriately provided with a bulbous head ornamented with a spiky helmet. The head

Figure 4. An early worm in drag. This sketch by the author shows a reconstruction of the fossil *Canadia*, found in the Middle Cambrian Burgess Shale of British Columbia in Canada, which is some 505 million years old. This fossil is typically just 2–3 cm long. It is here shown as a polychaete worm, complete with simple eyes and bunches of bristles. *Canadia* is among the oldest examples of the Phylum Annelida—which includes the earthworms and the ragworms—to appear in the fossil record.

then passes back into a long wrinkled body and the whole was then protected by some kind of rubbery organic sheath. According to Simon Conway Morris of Cambridge—who, amongst much else, is an acknowledged expert on fossilized penis worms—*Paraselkirkia* was not an annelid but a priapulid, a penis worm.

Indeed, we must here admit that—*Canadia* aside—many Chengjiang creatures closely resemble forms from the younger Burgess Shale biota of Canada. The latter even shares with Chengjiang a little nightmare of a beast called *Hallucigenia.* Many of us had observed this famous Burgess Shale fossil climb through the charts from folk lore to fame during the 1970s. Each Christmas, at the Palaeontological Association meeting in England, we would sit down to receive a batch of new revelations made by the group of Harry Whittington at Cambridge. There was even some unkind speculation that his group had made a Faustian pact with Pan in order to be able to discover extraordinary new phyla at such a rate.

Whatever the pact, its almost Faustian trajectory was wondrous to behold. All of this was to culminate in a claim that some now call 'the Great Burgess Shale Bubble'. That claim was mainly written up within a famous book called '*Wonderful Life*'.[51] It is a book that can still be read with profit, especially to naughty children at bed time. In it, that great author Stephen Jay Gould lovingly describes a battalion of Cambrian monsters, not all of which hailed from Cambridge.

What Gould was saying was this: the extreme range of body plans across the animal groups was much greater in the Cambrian than it is now, and that the accidents of time have progressively whittled down this variety, so that all we have now is a few lucky survivors. In other words, our own ancestors were not so much 'fit' as very lucky indeed.[52] But extraordinary claims require extraordinary evidence and Gould, it seems, was arguably falling into a grave error. That old Burgess bubble was about to burst, even as he wrote. One fossil, in particular, was to help burst that bubble in the very messiest of ways—a little fossil called *Hallucigenia.*

It is instructive to take a careful look at *Hallucigenia*, to see how thinking about it has itself evolved over the past century or so. This fossil looks rather like a caterpillar that has been squashed very firmly under a Wellington boot on a wet garden path. Its elongate body is divided into segments, and provided with seven pairs of fleshy legs. And like the caterpillar of a modern Silkworm moth, for example, it had almost as many pairs of spiny nobbles along its back as it had legs beneath. Like those of a caterpillar, too, these spines may have helped to deter any would-be predators. All that seems rather obvious to us now, at least in the excellently preserved Chengiang material. But, unhappily, it was not so easily seen in the less than well-preserved Burgess Shale material of the very same form. Thus it was, back in 1911, that the American geologist Charles Walcott had mistakenly regarded *Hallucigenia* as a kind of polychaete worm like *Canadia*, with spines for bristles. And so it stayed, just another wormy drag queen, unloved and unremarked, and accordingly tucked away in the drawers of the Royal Ontario Museum of Toronto and the Smithsonian Institution of Washington. There it slept until it was awakened in the 1970s.

The 1970s was the decade in which hippies were being replaced by heavy metal and punk rock. *Hallucigenia* was duly given a hippy name and a punk make-over. Simon Conway Morris at Cambridge whimsically reconstructed *Hallucigenia* both upside down and back to front. Unhappily, it was seemingly made to walk with stilts on its back and feed via openings in its little wiggly legs, presumably passing food for ingestion through its bottom and then excreting faeces through its head. Thus rendered, *Hallucigenia* was no longer an old drag queen. She became a punk tour-de-force, all fossilized spikes and safety pins. Fortunately for the post-punk generation, a Swedish palaeontologist called Lars Ramskold saved the world from this hellish vision. We learned that *Hallucigenia* was nothing more than a velvet worm—a bit like the modern *Peripatus*—but provided with dainty spines on its back and with little legs to dance upon. Its dance had changed, as well, from punk rock to a peripatetic rumba in just a few weeks. *Halucigenia* was then to be called, rather unkindly, the lie in *Wonderful Life*.

Other velvet worms then started to turn up in the Chengjiang biota, including the goggle-eyed *Microdictyon*, whose *Hallucigenia*-like spines were converted into huge false compound eyes—perhaps an example of mimicry to warn off any would-be predators (see Figure 5). Following these amazing discoveries, a reaction began to set in. The fossils of the Burgess Shale weren't odd, we learned. It was the reconstructions that were odd—as was the philosophy of their reconstruction, if such it could be called. While Gould was dreaming about *Hallucigenia*, many were watching with morbid fascination as the Burgess Bubble began to burst. One by one, those weird and wonderful Burgess Shale fossils started to claw their way back into the fold of more familiar animals. Richard Fortey quipped with us during that time—'there is nothing as odd as a brine shrimp or as balmy as a barnacle in the Burgess Shale biota. It's just a load of old codswollop.' Or words to that effect.[53]

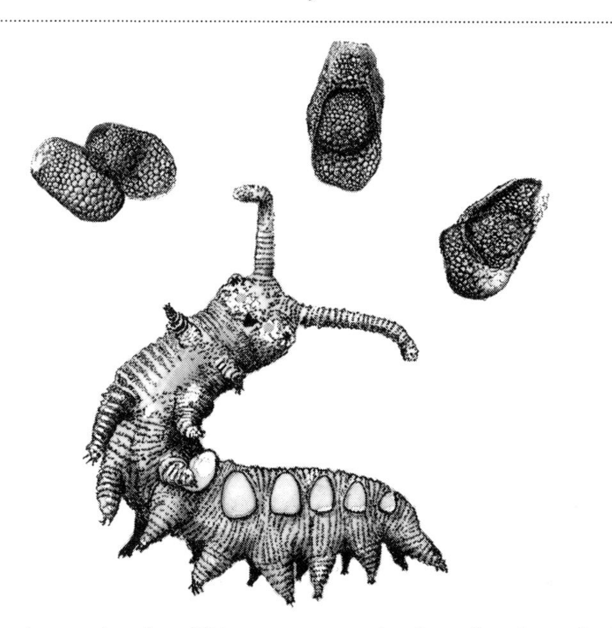

Figure 5. A worm in velvet. This reconstruction by the author shows the fossil *Microdictyon*, found in the early Cambrian Chengjiang biota of south China. This fossil, which is about 520 million years old, is typically just 2–3 cm long. Such fossils are usually preserved as isolated phosphatic plates, a few millimetres across, each looking a bit like a trilobite's eye. In the Chengjiang biota, these phosphatic plates are also found in pairs (as shown above) and comprise the dorsal armour of a creature that can be reconstructed as a caterpillar-like velvet worm (drawn below). *Hallucigenia* had simpler and more pointed plates than *Microdictyon*. Both are among the oldest known examples of velvet worms in the Phylum Lobopoda.

A Red Gauntlet is thrown

The Burgess Shale biota is historically important for us. But the Chengjiang biota is scientifically more so, because it is some 20 million years older. As such, it provides us with a rich menagerie of animals from the time of the very earliest trilobites. But its important message was, and still is, not wholly about Cambrian

Figure 6. The Meishucun phosphate mine. These richly fossiliferous phosphate quarries were shown to the author by the quarry manager (at left), and palaeontologists Yue Zhao (middle) and Jiang Zhiwen (at right). The first shelly fossils, including *Anabarites* and *Protohertzina*, are some 540 million years old and appear at the level of their feet, while molluscan shells related to *Maikhanella* and *Aldanella*, appear shortly above their heads.

weirdness at all. Both the Burgess Shale and the Chengiang fossils show us something equally peculiar—that Darwin's old dictum '*No organism wholly soft can be preserved*' cannot easily be applied to the strange world of the Cambrian.

Chengjiang also confirms that Cambrian oceans swarmed with creatures that were arguably rather close to the direct ancestors of our modern marine and terrestrial animals and, of course, to ourselves. But where did all these creatures come from? And did they really arrive so suddenly? To follow this question, it is time return to the Country House murder mystery we attended in

Chapter 2—in other words, to the meeting of palaeontologists at Burwalls in 1983.

Few of us in the West had been aware of the frenzy of geological mapping that had taken place in China during the 1970s, after the Cultural Revolution. Searching for fertilizer in the form of phosphate rock, geologists had accidentally stumbled upon a rich vein of ancient fossils that could be traced for thousands of kilometres across southern China (see Figure 6). An orgy of fossil description then followed. According to one jibe, there were more Chinese palaeontologists naming fossils from the Precambrian–Cambrian transition than there were fossils for them to name. Happily, this was to prove untrue.

Back in the meeting at Burwalls in 1983, the Soviet presentations had met with glows of approbation or scowls of scorn. That hubbub soon gave way, though, to a hush of anticipation as four Chinese scientists took to the podium. A red gauntlet was about to be thrown down—by unfamiliar geologists who were among the first to set foot outside Red China. We leant forward expectantly to hear what the Chinese scientists had to say.

The Chinese interpreter began by apologizing profusely for his poverty of spoken English. He then went on to talk in immaculate language, complete with jokes and asides. Here was no ordinary 'translator'. It was followed by an impassioned account—from Drs Xing Yusheng, Luo Hulin, and Jiang Zhiwen—of surprising discoveries that were starting to be made in China. The Tommotian fossils of Dr Alexei Rozanov, we heard, were far from being the oldest evidence for animals with skeletons. Chinese geologists were finding an even lower, and hence older, assemblage of fossils, which they called Meishucunian, named after the quarries at Meishucun in which they were found. Those rocks contained *no* evidence for brachiopods, snails, or sponges like those seen at Ulakhan-Sulugur and Dvortsy in Siberia. Instead, Chinese researchers were finding rather obscure and extinct fossil groups, many of them tubular and sickle-shaped.

Understandably, the feelings between Russians and Chinese began to run a little high after this pronouncement. The Moscow view seemed to be that these Chinese rocks were the same age as their own Tommotian rocks, simply of different aspect.[54] To the Chinese, however, their 'small shelly fossils' were respectably older than the Tommotian of Siberia and therefore had greater importance in terms of evolution.[55] To help resolve this question, at least in my own mind, it was necessary to travel to China.

Bamboo Temple

By September 1986, I was able to spend about a month, as a guest of the Geology Museum in Beijing.[56] The Cultural Revolution was over but its scars were everywhere.[57] No surprise, then that the Geology Museum in Beijing was filled with a huge but seemingly forgotten collection of fossils and minerals. There were no tourists to speak of, and I remember no school children or other visitors. The mood was sombre, too. During the Cultural Revolution, even the senior staff with whom I was working had been forced to labour in the paddy fields.

At the start of each day, I was picked up by a chauffeur driving an 'official car', that is to say, it had curtains and he was wearing white gloves.[58] I would be met at the door of the Museum by my translator Mrs Zhang. We would trail down long green corridors which rang out, every now and then, with a shudder-inducing *ping* from one of the many giant spittoons arranged along the walls. Evenings were equally memorable. Dinners called 'banquets' were punctuated by speeches and spirited toasts from each of us around the table.

After visiting this vast Museum, I was driven around south China in an old black limousine, with palaeontologist Yue Zhao as my guide. For day after day, we peered at the strange world outside through car windows provided with dainty net curtains. For mile after mile, our chauffeur-driven vehicle was the only car on a road packed with pedestrians, bicycles, and ox carts. Every now and then, we would stop to negotiate potholes or landslides. Small crowds would gather to stare at me solemnly, perhaps their first ever sighting of what they call a 'long nose'—a European. I felt honoured. At one place, we were ushered over to a great wooden platform that hung over the field below, and urged to make 'deliveries' into the ox cart that stood beneath, ready for transport to the paddy fields. It would have been rude not to oblige.

Gradually, we pushed further and deeper into China, finally reaching the 'perpetual spring city' of Kunming. Disconcertingly, a thin pall of smoke hung over the tower blocks and seeped across public gardens that fringed the edge of town. Even so, these gardens, with their willow trees, orange blossom, pagodas and little bamboo bridges, seemed to come straight from an old willow pattern plate. Each day for a week, we would drive out of town, past the gardens, and into the surrounding hills scattered with the remains of Buddhist temples. Monks were barely to be seen there, of course. Those places of worship felt then like empty husks, mere museums with idols, bells, and fountains.

One of these, the Bamboo Temple, lay on a track up to Qiongzhusi, which was one of the first places to bring forth the famous Chengjiang fossils. The temple dated back at least to the seventh century AD, but the previous year it had burnt down in a fire. In the courtyard, before a large pink Buddha, sat a miniscule monk collapsed in a low chair with eyes tight shut. Every few minutes, though, he would leap up and tap the large bronze bowl beside him with a wooden hammer, so that it chimed like a town clock. Peering closely at the 'gong', we could see that it was actually filled with rubbish, mainly old newspapers, to ease his eardrums. Beyond this precinct, with its dragons and gongs, we came across a lurid row of life-sized figurines. Gods and demons were standing shoulder-to-shoulder in a frieze, showing moods from angry to sublime. But what surprised me most, when we turned a corner, was a little museum. Within a series of picture frames squatted a dozen little fossils collected from the local hillside. Each was labelled in Chinese characters: *Eoredlichia*, 名称 古莱得利基虫. What a joyous thing to see—fossils in a Buddhist temple! And why not? It made me think. Would any Evangelical Church dare to celebrate the history of life in this way?

Wangjiawan

A good place to visit the Cambrian explosion in China, we soon learned, was to be found in the phosphate quarries near Meishucun.[59] But in China in those days, such phosphate mines were a little unhealthy and dangerous, being something like working prisons. We discovered a much finer setting along the remote mountain hillside of Wangjiawan, south of the picturesque Dianchi Lake. To reach it, we drove round the lake so that we could hike in from the road. Anxious to get there in good time, our guide, Jiang Zhiwen set off in front at a cracking walking pace.

After a hike of about 4 km, we reached the object of our quest, the phosphate rock of Wangjiawan. With help from the fossils, we attempted to find the time zone represented by the base of the Tommotian at Ulakhan Sulugur. As in Siberia, this evolutionary turning point seemed to be marked by a dramatic change in rock colour. But here the change was from brown below to black above—almost like jet—rather than from white to red as in Siberia. This section at Wangjiawan also showed us something that we could not see along the River Aldan. The black-coloured, Chinese 'Tommotian' was here underlain by beds of brown phosphate up to ten metres thick. And it was richly fossiliferous.

This layer of shelly fossils within the phosphate beds at Wangjiawan was an important Chinese discovery because it confirmed, for many of us, the presence of diverse animal life before the Tommotian. But none of the modern animal groups, like brachiopods and echinoderms, could clearly be recognized among the shells at this lower level. Instead, the Chinese palaeontologists were finding an odd collection of unfamiliar looking shells—strange tubes, tiny teeth, and spotted spheres—just as they had said at the meeting, in fact.

One of the most characteristic fossils of these Meishucun beds is called *Anabarites* (see Figure 7). It was first described from rocks around the Anabar River in Siberia, far to the north, and has been found near to the very base of the Cambrian right across the planet, from the Rocky Mountains in Canada to the foothills of the Himalayas in India, and here again in China. The ancient shell of *Anabarites* is tubular, like a tiny drinking straw, and seldom more than a centimetre or so in length. A most curious feature is its threefold symmetry, giving it a cross-section like a clover leaf, and producing three long ridges that run down the sides. When looked at in cross section, we can bisect it along three different planes along which one half of the fossil can be reflected in a mirror to reconstruct the whole beast. In other words, the shell has three reflectional symmetries. Likewise, the shell can be rotated through three positions in which it will still look the same. That means it has three rotational symmetries. Curiously, there is no shell with a symmetry pattern quite like *Anabarites* still living today, nor even

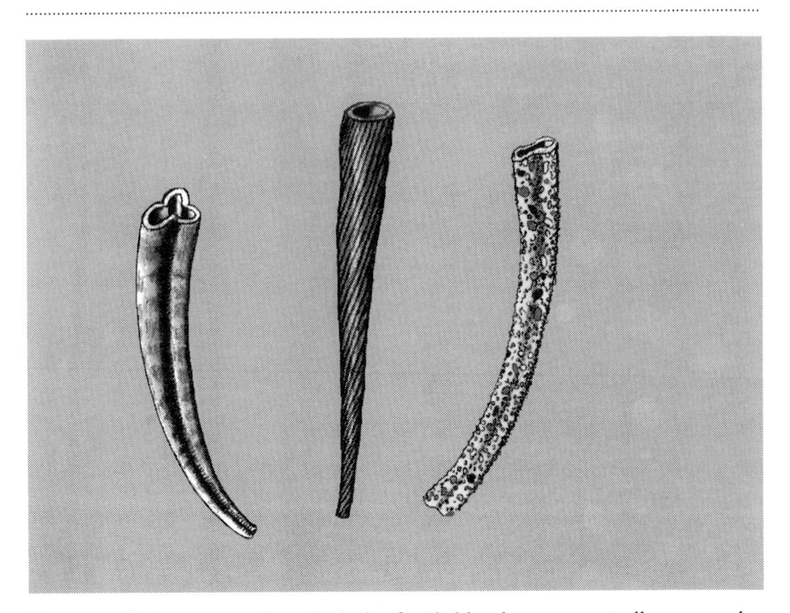

Figure 7. Tubeworms galore. Tubular fossils like these are typically among the first shelly fossils to appear in the fossil record, and flourished well before the appearance of trilobites. *Anabarites* (at left) had a shell of chalky material provided with three longitudinal grooves; this may be a relative of the Phylum Cnidaria. *Coleoloides* (at centre) had a chalky shell provided with numerous spiralling grooves; this may have been made by an invertebrate worm. *Platysolenites* (at right) had a shell constructed from small mineral grains glued together, and was almost certainly made by a single-celled foraminiferid protozoan. Each of these fossils is seldom more than a few centimetres long.

in the rest of the fossil record. This style of shell seems to have been fashionable in the earliest days of the Cambrian explosion, only to become outmoded once the trilobites, like *Fallotaspis* and *Eoredlichia,* had started to flourish on the seafloor.

We still know surprisingly little about the ecology of these earliest shelly fossils. One of the problems is that the earliest skeletons, like those of *Anabarites,* occur in shell beds that have been tumbled about in storms. That makes it difficult to imagine its original way of life. Even so, the long tubular shell provides some useful clues. Such tubular shells are especially popular today among creatures that wish to raise themselves above the sea bed to feed on organic materials suspended in the water column, and are appropriately called suspension feeders. A classic example of such suspension feeding can be found in the reefs we have seen around Barbuda, where a snail called *Vermetus* attaches to the coral and then 'uncoils' into a long tube that extends outwards into the swell. Like a garden slug, *Vermetus* can produce prodigious amounts of slime, which it dribbles into long streamers that trail outwards with the current column. A few hours later, this snail painstakingly hauls back the trail, eating the snotty streamer plus its entrained food particles. Like a jam tart dropped unwittingly on the kitchen floor, it usually picks up more than it loses. Disgusting but effective. Indeed, this strategy can be so successful that great meadows of vermetid tubes thrive in many places where the ocean waters are rich in food. From this observation, we can build a hypothesis that *Anabarites* probably lived attached to rocks in meadow-like colonies, feeding from a water column enriched in food particles. The close association between *Anabarites* fossils and phosphate rock—which requires lots of nutrients to form—also makes this hypothesis appealing.

Curiously, wherever we have looked, the earliest communities of shelly animals seem to have been of this kind—great meadows of tube worms. We still have no idea whether these 'tube worms' fed using strings of slime or used some other means of entrapment, such as 'feather dusters' or tentacles. Nor are we yet sure what kinds of animal lived inside each tube. After all, a tube is just a tube. It gives away few secrets about growth and biology. Most agree, however, that *Anabarites* and its relatives were made by some kind of invertebrate (perhaps an annelid worm, a jellyfish or a coral) but that is about as far as we can yet go. Of one thing we can be sure, though: meadows of tubular creatures seem to have been all over the place, slurping up a rich soup of food particles at the start of the Cambrian.

A fossilized jelly baby

When Darwin wrote '*No organism wholly soft can be preserved*'[60] he had not reckoned with the strangeness of the fossil record, nor yet with the vast explanatory power of his Big Idea. Life is not the only thing that has evolved. *Even fossilization has 'evolved'.* The further back in time we go, as we shall see, the quality of the very best preservation seems to get better and better. How very odd that sounds. It seems to be the very opposite of what we, as well as Darwin and Lyell, had been expecting.

During the age of the dinosaurs, for example, cases of remarkable preservation are rather rare and largely restricted to the beds of ancient lakes. But much earlier, around 520 million years ago, at the time of the Chengjiang biota with its early trilobites, examples of soft tissue preservation on the seafloor seemingly abound. Such soft tissue preservation means, of course, looking at gruesome dead bodies. Happily, the more gruesome the cadaver, the better we palaeontologists like it. Squishy creatures with disgusting innards intact have obligingly turned up—to our evident delight—in marine mudstones from Utah to Canada, from Greenland and Russia, and from China and Australia.

At this level where we now stand in China, near the starting line of the Cambrian explosion, the quality and abundance of remarkable preservation can be awesome. So extensive was this phosphate bed, that it is now being mined for fertilizer across southern China and into India, Pakistan, and Kazakhstan.[61] These earliest remarkable fossils are so abundant in the rock that they are innocently ground up into powder for use on the fields as fertilizer. Common as muck.

To stumble upon this curious revelation, I had travelled westwards from Yunnan into Sichuan to join He Tinggui at the College of Geology in Chengdu. The campus was like a little city, with its own football stadium, theatres, and cinemas, for the training of a mere 1700 geology students per year. In 1986, it was one of seven such colleges in China. When Yue Zhao joined me for a walk around the campus one evening, we mingled with the newly arrived freshmen students who were buying rush mats, carrying bedrolls, purchasing dumplings, soling shoes, and carting their baggage about in trishaws—tricycle carts. Curiously, there were no cars, newspapers, or televisions. News broadcasts instead took the form of big sheets of paper pasted to the walls, often bearing clips from the Thoughts of Chairman Mao. Crowds of students were gathered around these poster boards. But the lack of news didn't make for peace and quiet in Chengdu College. We were woken each day by shrill and echoing announcements over the electric loud hailer system.

From the packed College of Chengdu, we began our quest by journeying some 170 km, westwards yet again, to the Sacred Mountain—called Emei Shan—where the Cambrian fossils lay. We had come to collect from the famous basal Cambrian phosphate bed, balanced precariously on the flanks of a steep mountain slope. As Yue Xhao and He Tinggui set about filling the sample bags with phosphatic rock, I paused to gaze in wonder at the scenery (see Plate 5). Serried ranks of conical hills, draped in lush green forest, would disappear from time to time beneath veils of mist that descended from the sacred mountain slopes above. Tiny terraces, all neatly tilled, clung to the hillsides. As we worked, Asia herself flowed past—maidens laden with digging tools; an old lady being carried downhill on a pole; a black butterfly the size of a bird, two men carrying a trussed pig to market. Sichuan can seem like the busiest place on the planet.

Back in the lab next morning, we looked down the microscope at the kinds of fossil we had been collecting. There indeed, was *Anabarites*, exactly as expected. But one sample, from the middle of the series, showed something like fossilized Chop Suey—tangles of noodles interspersed with mushy peas and cashew nuts, peppered through with little morsels of meat, all preserved in mineral phosphate. These fossilized noodles were the remains of filamentous seaweed from the Cambrian seafloor. The mushy peas and cashew nuts were pimply spheres called *Olivooides*, believed at one time to be algal resting stages. And those morsels of meat were thought to be the remains of long extinct jellyfish, close to *Arthrochites*. The sight was entrancing. Over the coming years, both Yue Xhao from China and Stefan Bengtson from Sweden were to show that these mushy peas were probably the baby growth stages of the meaty morsels—embryo, foetus, and adult—all preserved under strange conditions on the seafloor.[62] Intriguingly, such conditions seem to have been widespread at the very start of the Cambrian, spreading across Asia, Europe, Australia, and North America. Here again, was evidence that Darwin's great prophecy: '*No organism wholly soft can be preserved*' did not apply to this Lost World. What on Earth was going on?

The Guessing Game

This research coming out of China started to reinforce a pattern that was beginning to emerge from the fossil record. There was indeed a rich world of animals with skeletons before the appearance of trilobites like *Fallotaspis* and *Eoredlichia*. Chinese evidence also confirmed the impression of a Cambrian explosion that was not so much instantaneous as spread over tens of millions of years. Most importantly, Chinese work started to amplify the muffled words of Russian scientists working in Novosibirsk concerning the concept of a pre-Tommotian world lacking brachiopods, snails, or archaeocyath sponges. That ancient world had seemingly been dominated by tube worms such as *Anabarites*.

Our knowledge of the fossil record was therefore growing to the point where we could start to decode *patterns* in the evolution of a particular animal group. Seeing such patterns should therefore make it increasingly possible to hazard some guesses—that is, to formulate some hypotheses—about the name of the game being played. Remember, though, *we are never told what the rules are with the fossil record.* Our great aim must be to gain a better understanding of this game and its in-house rules. It would be wise to remember, therefore, that all initial guesses are likely to fall somewhat wide of the mark.

Here then, comes our big question again: *was the apparent lack of animal fossils in the Precambrian—Darwin's Lost World—a real phenomenon or was it a kind of 'bluff'?* In terms of our game of cards, were the missing cards still in the deck (hidden in the rock record), in the hands of other players (perhaps hidden away in museums), or simply not there at all? Historically, three main gambits have been put forward to flush out an answer to this seminal question. We will name these gambits after three original thinkers: Lyell's Hunch, Daly's Ploy, and Sollas's Gambit.

Lyell's Hunch

In the middle of the nineteenth century, the conventional view about the Cambrian explosion was essentially a creationist one.

We can see this from some of the words that Darwin wrote in 1859: 'To the question why we do not find records of these vast primordial periods, I can give no satisfactory answer. Several of the most eminent geologists, with Sir R. Murchison at their head, are convinced that we see in the organic remains of the lowest Silurian stratum the dawn of life on this planet'.[63] But Charles Lyell had a different explanation. He blamed the absence of skeletal fossils below the Cambrian on an inferred incompleteness of the fossil record. Darwin followed in Lyell's footsteps here, and suggested that the oldest fossils had all been hidden and destroyed by burial deep below the modern oceans. In Edwardian times, the great American geologist Charles Doolittle Walcott went one step further and conjured up a world in which animal ancestors had been hiding away, through the later Precambrian period, in a lost 'Lipalian ocean' whose deposits had yet to be found or had disappeared. Both visions belong to the view that we shall here call Lyell's Hunch, because Sir Charles Lyell—Oxford's most famous geologist—thought of it first.

To gain a clear feeling for Lyell's Hunch, we cannot do better than receive it from Lyell's own pen, here as written in 1853:

> If we next turn to the fossils of the animal kingdom, we may inquire whether, when they are arranged by the geologist in a chronological series, they imply that beings of a more highly developed structure and greater intelligence entered upon the earth at successive epochs, those of the simplest organization being the first created, and those more highly organized being the last. Our knowledge of the Silurian [i.e. Cambrian to Silurian] fauna is at present derived entirely from rocks of marine origin, no fresh-water strata of such high antiquity having yet been met with. The fossils, however, of these ancient rocks at once reduce the theory of progressive development to

> within very narrow limits, for already they comprise a very full representation of radiata, mollusca, and articulates [i.e. corals, molluscs, and arthropods] proper to the sea. Some naturalists have assumed that the earliest fauna was exclusively marine, because we have not yet found a single helix, insect, bird, terrestrial reptile or mammifer; but when we carry back our investigations to a period so remote from the present, we ought not to be surprised if the only accessible strata should be limited to deposits formed far from land, because the ocean probably occupied, then as now, the greater part of the earth's surface.[64]

'Lyell's Hunch' is here taken to refer to the idea that the Cambrian explosion is not real—the Precambrian teemed with animals as yet unfound. It was arguably a necessary step—an unwillingness to accept negative evidence or any kind of predestination in the history of life. It was effectively a null hypothesis against which to test the mounting claims of 'progressive development' in the history of life. Until we have evidence to the contrary, Lyell was saying, then we should assume that everything in the past worked in exactly the same way as we see now—the famous Principle of Uniformity. It was, and still is, a very strong hypothesis for making powerful predictions.

Darwin must have been well aware of some of the dangers in using Lyell's Hunch back in 1859. The problem for him was that the fossil record was still too poorly known to allow the story to move forward. When writing the *Origin of Species*, he must have felt the pressure to guess 'the name of the game', as we have seen. We can only surmise what he was really expecting the fossil record to show. But it cannot have been too far removed from the idea that animals, as we currently understand them, only enter the fossil record with the Cambrian explosion but that there was a long and rather gradual progression towards them in the distant past.

This hunch is still popular with some molecular biologists today, especially when attempting to reconstruct evidence from the genome of living animals. Such reconstructions are sometimes done in isolation from the fossil record. Fair enough, it can be said. This isolation from fossil evidence has the seeming advantage of avoiding negative evidence which, in this case, is an absence of fossils or of molecules. And negative evidence is sometimes regarded as weaker than positive evidence in any debate because—as astrobiologist Carl Sagan used to say—absence of evidence is not evidence of absence. This dictum can be illustrated, for example, by the so-called Lazarus fossils. These are creatures that have seemingly arisen from the graveyard of the fossil record, most famous of which is the *Coelacanth*. This primitive scaly fish was believed to have died out with the dinosaurs at the end of the Cretaceous until, that is, it was caught alive by fisherman off the coast of Madagascar.[65] Less well known, but equally remarkable is the case of a primitive shellfish called *Neopilina*. This comes from a group of molluscs that were thought to have become extinct some 300 million years ago, until it was dredged alive from the deep sea off Puerto Rico. Equally romantic is the story of the Maidenhair tree, *Ginkgo*, which was first described from the fossil record, only to emerge alive and well in a monastery garden in Japan.

The dismissal of negative evidence may seem fine during the early stages of research.[66] But negative evidence is an integral part of the mathematics of pattern and probability. The longer any game is played, the clearer must become the patterns. Geologists, for example, have sat down and played the game of Lyell's Hunch for nigh on 200 years or so, and yet—as we shall see—no convincing evidence for trilobites or higher animals has ever turned up in the Precambrian. Let us not forget, either, that poor old Lyell was half-expecting fossil mammals to come tumbling out of Palaeozoic rocks.[67]

When asked what single observation would disprove evolution, Oxford biologist J. B. S. Haldane famously growled: 'fossil rabbits in the Precambrian!'[68] Those long-awaited reports of cats in the Cambrian and of rabbits in the Precambrian never arrived, though. Even by 1859, Lyell's Hunch was therefore starting to seem just a little bit shaky. The predictive power of this gambit was never fulfilled. And when the predictions of a gambit begin to fail, its followers will start to falter and its use will decline.

We can now see that Darwin, not wishing to inflame the Establishment, studiously avoided all talk about 'progressive development' and opted instead for a long diatribe on the imperfections of the fossil record. As we shall see, however, his prognosis on this was far too gloomy. He clearly felt obliged to back a hypothesis that was essentially *non-evolutionary*. Darwin was trapped in a hypothetical web of Lyell's making. But the puzzle of Darwin's Lost World was soon to be interpreted in other ways.

Daly's Ploy

Charles Darwin told his friends that the lectures he attended as a student in Edinburgh given by mineralogist Robert Jameson were woefully tedious. Poor Jameson was by that time regarded as quaintly antediluvian because he was the disciple of a famous eighteenth-century mineralogist called Abraham Werner. The latter was the first scientist, as far as we know, to ever give a course of lectures on fossils, at Freiberg in 1797. But it is Werner's views on crystalline rocks, like granite and gneiss, that we need to consider here. It was Robert Jameson's view, and Werner's mantra, that the crystalline rocks were laid down in the early stages of evaporation from a great primordial ocean. Superimposed layers in the rock record, like those we have seen in Siberia—dolomite below limestone below shale—could therefore be read as a reflection of the changing chemistry of the oceans.

Werner's hypothesis is both beautifully platonic and sadly flawed. Two colleagues of Robert Jameson—James Hutton and John Playfair—were able to argue against him by showing, in the hills around Edinburgh, that most crystalline rocks were not sediments at all but were of igneous—volcanic—origin.[69] This debate, which was called the Neptunist–Plutonist controversy, took place at the height of the great Scottish Enlightenment. At the end of the debate, the Neptunists were obliged to cave in. Even so, Werner deserves our kindly remembrance for his contributions towards two important concepts. First, that the layers of rock represent successive events whose story can be read.[70] And secondly, that the chemistry of the ocean has changed—we might now say it has evolved—during the vast expanse of time since things began. In other words, Jameson was, almost by definition, on the opposite side of the fence from Lyell and his Principle of Uniformity.

Werner of Freiberg is of interest to us here because he argued that the primitive oceans had a chemistry hostile to animal life before the start of the Cambrian. This was a highly influential idea that gained support, in modified form, from many later writers including Sir Charles Lyell and Robert Chambers. Even in 1859, it was still widely believed that crystalline rocks such as granite, gneiss, and schist were the remains of an early ocean with temperatures reaching to the boiling point of water.[71]

The late Victorian geologist R. A. Daly resurrected some aspects of this Wernerian viewpoint, suggesting that the early oceans before the Cambrian were without carbonate of lime (calcium carbonate), meaning that animals could not secrete the hard shells needed for preservation in the fossil record.[72] But a problem with this idea quickly emerged—the early rock record is rich in lime—miles upon miles of limestones and dolomites—as we shall later see.

A more recent version of Daly's Ploy has been put forward by American geologist Paul Knauth, who has suggested there was far too much salt in the early oceans. In this view, the earliest plants and animals evolved in freshwater lakes. The impoverished tally of animals from the salt ponds of Cuffy Creek in Barbuda, and their contrast with the rich diversity of animals in the normal marine salinities of Goat Reef (see Chapter 1) shows us what might be expected from such a scenario.[73] But most of the creatures we met in the Cambrian—brachiopods, sponges, echinoderms–abhor freshwater today and, so far as we can tell, in the past as well.

All versions of this idea—that we shall here lump together under 'Daly's Ploy'—predict that environmental controls—that is to say, extrinsic factors beyond biology—will turn out to have been the cause of Darwin's Lost World and its nemesis, the Cambrian explosion. These 'extrinsic factors beyond biology' have been taken to include heat, hydrostatic pressure, poisonous gas, salt, or calcium, or phosphorus, or oxygen. Or several of these. Or even all of these. Whereas Lyell's Hunch assumes that the rules of the game are both fixed and simple, and that we have simply missed a trick, Daly's Ploy says that the rules are neither fixed nor simple. They can change, and change in ways that might seem 'unfair'. Remember, *we are never told what the rules are with the fossil record.*

Sollas's Gambit

It is with the next gambit in the game that we meet with 'Sir David and the Gully Quartzite'. The 'David' here is Sir Edgeworth David. He was a famous Oxford geologist who went on to become a great explorer, striding alongside Sir Ernest Shackleton in Antarctica. But he also camped out in the bush of South Australia, episodically from 1896 to 1926, in search of soft-bodied Precambrian animals from the Gully Quartzite, at levels far below the Cambrian trilobites. He deserves credit as one of the first scientists to explore this question of Darwin's Lost World in a systematic way.

David drew attention to the fact that the shells of animals tend to get thinner as one travels back down the fossil record from the Silurian, with its thick-shelled molluscs and corals, back to the Cambrian with its thin-shelled creatures like *Lingulella*. Taking this line of thought, he predicted that animals before the Cambrian may have been naked, with no shells at all.[74]

It might seem that Sir Edgeworth David was referring to something like Daly's limeless ocean. But he had clearly mapped Precambrian limestones right down at the level of the Gulley Quartzite. He even named them—the Torrens Limestone, the Brighton Limestone, and the evocatively labelled Blue Metal Limestone. So he knew the Precambrian ocean could not have been limeless. Not only that, but he had also worked with his old Oxford colleague—William Sollas—on drilling through limestones on Funafuti atoll in the middle of the Pacific Ocean. So David knew all about limey skeletons as well.

Happily for us, William Sollas explored the same big idea, because it was seemingly his own idea. Here is how he put it, rather clearly, back in 1905:

> How is it that, with the exception of some few species found in beds immediately underlying the Cambrian, these have left behind no vestige of their existence? The explanation does not lie in the nature of the sediments, which are not unfitted for the preservation of fossils [i.e. Lyell's Hunch is wrong], nor in the composition of the then existing sea water, which may have contained quite as much calcium carbonate as occurs in our present oceans [i.e. Daly's Ploy is wrong]; and the only plausible supposition would appear to be that the organisms of that time had not passed beyond the stage now represented by the larvae of existing invertebrata, and consequently were either unprovided with skeletons, or at all events with skeletons durable for preservation.[75]

In other words, Sollas was saying that we are seeing in the Cambrian explosion the effect of the evolution of easily fossilized mineral skeletons.

William Sollas was one of the last great polymaths of his age. At one and the same time he was a world expert on sponge biology, on freshwater ecology, on hominid evolution, and on the great age of the Earth. He was even accused by a colleague, of somewhat more modest ability, of being the true Piltdown hoaxer.[76] All this scope gave to him—as it gave to Darwin—a remarkable sagacity and perspective, especially so in mid-life. Sollas was therefore to come close, very close indeed, to guessing the nature of the cards in the game. But *pattern* is not the same thing as *process*. There were therefore two big questions that had still to be explained: Was the Cambrian explosion real, in terms of the evolution of major animal groups, and if so why? And was there truly a change from naked to skeletal animals, and if so why?

Sollas could have had no hope of answering these questions back in 1905. That is because key clues were still unfound. Indeed, they have only recently come to hand. To discover what these clues might be, we must leave China behind, and travel over the mountains and across the Gobi Desert, to arrive in the land of Genghis Khan—in deepest Outer Mongolia.

CHAPTER FOUR

THE FIRST TERROR WITH TEETH

A steppe in the right direction

Standing outside my circular tent, called a *gher*, on a windy afternoon in August 1993, I felt like pinching myself to check that life was real. Earlier in the day, the sky had been peppered with eagles circling above our little cluster of nomadic ghers, pitched by local tribesmen from the Dzabkan River. Camped up here, on stony ground in the Gobi-Altai Mountains, not far to the north of the Gobi Desert, I scanned the horizon for signs of Mongolian horsemen. The air was crystal clear but I could, as yet, see neither horses nor men. Nor could I see trees or fields. In fact, I had seen neither fields nor vegetables for several weeks. Here I was, then, an ancient fossil hunter in an ancient hunters' landscape. Grass, mountains, rivers, and snow. It felt as though I had travelled back in time some twenty thousand years, to an age when our ancestors had painted the ceilings of caves in distant Lascaux and Rouffignac.

A group of fellow geologists had gathered from around the world to join me in examining the outer Mongolian fossils (see Plates 6 and 7). We were sitting in the dining tent and had just eaten lunch. Careful dissection with a pocket knife showed it to comprise a kind of mutton, leavened with dollops of rice. Swilled down with a mug of hot china tea laced with mutton fat, it was a rather welcome addition to our modest calorie store after a chilly morning spent working on the rocks. But, it must be said that lunch that day had been slightly better than usual. On previous days, we had been served what seemed like a potage of sheep's anal sphincter jumbled together with goat's entrails. Only it didn't taste quite as nice as all that might sound. There was a sneaking suspicion that our Mongolian cooks—good folk—were dining just a little bit better than we were. After a week of this, one of the French scientists in our expedition decided to capitulate and switch camps, moving in with the cooks. Nobody could blame him.

Outer Mongolia is no place for the squeamish. And it is certainly no place for the squeamish vegetarian. Travel may, indeed, broaden the mind, but it can often narrow the stomach. The nomadic tribesmen hereabouts clearly dine on little more than mutton fat, yak milk, and horse flesh. Or, yet again, on yak fat and horse's milk. Or even on yak beer and horse's vodka. After a week or so, we began to dream of yak. And we began to dream, too, of Mongolian mountains piled high with potatoes and green beans. But there was ne'er a humble potato in sight, nor a green bean, banana, or lowly brussel sprout. One by one, we began to feel out-of-sorts. Had this diet continued over many more months, some of the team would no doubt have died or, more likely, walked home. Only one person went uncomplaining—my old mentor Roland Goldring. But Roland's culinary skills in the field were notorious. Not that mine are any better. But we all survived this Victorian regime of offal, fat, and rice. Interestingly, the average life expectancy hereabouts is as much as 60 years. Real Mongolians, it seems, burn off all that lard during the long cold winter months, or during much shorter bouts of wrestling and horse racing.

Frenzied wrestling and horse racing were, indeed, exactly what I was about to judge that afternoon. I say 'judge' because, as the grey-beard of our expedition, I was also expected to act as both judge and prize-giver for a racing and wrestling competition kindly put on for us by our hosts—Dr Dorjnamjaa and the local herdsmen. I had carefully constructed my profile so as to excuse me from taking part in these ephebic activities. But it did not excuse me from drinking jugfuls of fermented mare's milk, called arak—entirely ceremonial in nature, of course. This pale fluid looks a bit like washing-up water and it tasted faintly of yoghurt lightly scented with *eau-de-cheval-derrière*. It was certainly an acquired taste. But it was a taste I was obliged to acquire, since the alternative—refusing an honorific gift—would have been social suicide. And social suicide was something we dared not commit, marooned as we were, in the middle of the steppes of Outer Mongolia. The social niceties of Mongolia can matter very much indeed.

To assemble a story from the fossils of Outer Mongolia, I had attempted to draw together an unrivalled team of experts from around the world.[77] Over many weeks, this research team camped out upon the steppes, sometimes in ghers and sometimes in impoverished inns. Our arrival in this Gobi-Altai region was, to say the least, epic. Words can barely do justice to the dramas that we endured each day—vehicles ensnared in rushing mountain rivers, dire shortages of fuel, emergency nights spent on bare floor boards, a complete lack of potatoes. Happily, the lively good nature of Rachel Wood—whose rippling laughter could be heard for miles—spiced with the delightfully ironic wit of Simon Conway Morris, kept us all in good spirits. For most of the time.

We had come here, of course, not for arak but for a rock. And the sequence of rocks that we had come to see was reputedly better than anything we had yet studied in either Siberia or China.[78] Especially so, I think, in one important respect. The Mongolian story of Darwin's Lost World was found by us to be enormously expanded. That is to say, it could be traced down the mountainside and deep into the ravines, allowing us to step back in time through many thousands of metres of sediment, much of it replete with fossils. In other words, the rock successions hereabouts were much thicker, and therefore potentially much more complete, than those of similar age I had seen in most other parts of the world. We found that the fossil record in Mongolia extended downwards all the way down from the levels of trilobites, to the very earliest skeletal fossils. Indeed, the rocks just kept on going down and down, to the levels with pre-Cambrian animal fossils and thence to the Snowball glaciations. Not only that, but this peerless sequence of events could be traced across a considerable expanse of these foothills of the Gobi-Altai mountain ranges.

This hugely expanded sequence of rocks seemed to be just what Charles Darwin had been hoping for. And it could be explained, too, by a larger than usual rate of subsidence of the seafloor in this region at the time, much as he had been saying: 'The point in question is, whether widely extended formations, rich in fossils, and of sufficient thickness to last for a long period, would be formed except during periods of subsidence? My impression is that this has rarely been the case.'[79]

The comings and goings of the earliest animal fossils in these Mongolian rocks were to prove rather important for us.

They helped us to splice together a more detailed picture not only of the Cambrian explosion but also of its prelude, as we shall later see. Best of all, though, we had stumbled on lots of beds of limestone. That was good news because these calcareous beds can carry little chemical signals—like fingerprints—for testing both the relative age of the rocks and the conditions that existed on the ancient seafloor. These chemical signals would then allow us to correlate our old fossils not only across Mongolia and into Siberia, but also into China and from there into India, Iran, Oman, and, eventually, all around the world.[80]

The pearls

Many of the creatures we have so far met—such as *Anabarites* and *Chancelloria*—can look distinctly alien to the eyes of a modern biologist. But there is one fossil in Mongolia that would seem charmingly familiar to any conch collector: the tiny shell of one of the oldest known 'snails', called *Nomgoliella*—which is simply 'Mongolia' turned around, plus a bit on the end. On the seabed in the earliest Cambrian, the shell of *Nomgoliella* may well have shone with the iridescence of mother-of-pearl. A handful together could have looked like Captain Morgan's treasure chest. Unfortunately, in the rocks around the Dzabkhan river, the pearly shell of this little fossil was long ago replaced by calcium phosphate.

The shell of *Nomgoliella* has interesting mathematical properties too. It is constructed into a helical, logarithmic spiral. Like the shell of the modern Roman snail or Escargot, the spiral expands in a way that 'cuts any radius through the origin at the same angle'. First celebrated by Celtic artists way back in the Iron age, the magic of the logarithmic spiral was unveiled by René Descartes in 1638. But the most intriguing aspect of the logarithmic spiral seen in *Nomgoliella* concerns its different appearance when traced from place to place. The forms around the North Atlantic and Siberia are coiled in a right-handed way and are called *Aldanella* (see Figure 8). These examples from Mongolia are coiled in a left-handed way, and are called *Nomgoliella* or *Barskovia*. Specimens from China are barely helical at all and are called *Archaeospira*. And fossils from Iran have stretched out their right-handed coils to look like the horns of a mountain goat—these have no name at all. All these fossils are of similar age.

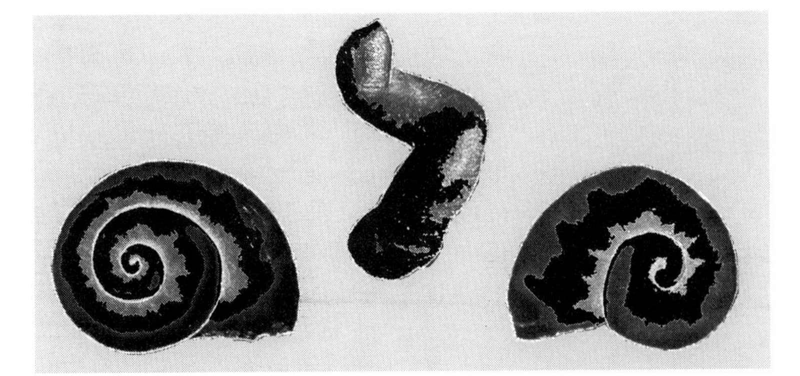

Figure 8. Shells with a twist. At left is *Aldanella*, a snail which coiled tightly in a right-handed way. This is the form typically found in Canada, England, and Siberia in rocks about 530 million years old. In the middle is an un-named form, which has an 'uncoiled' shell rather like a goat's horn. This is the form typically found in Iran. At right is *Nomgoliella*, which coils in a left-handed way. This is the form typically found in Mongolia. These little fossils are usually about 2–5 mm in diameter. They have here been reconstructed with markings like those found in the marine snails of Barbuda. These variations on a theme are all members of the Phylum Mollusca.

This conundrum is a bit like the paradox faced by Darwin when confronted with variations in the shell of the Galapagos tortoise from island to island. His explanation can also be applied to this example from the very beginnings of the Cambrian explosion: the original population had the potential for wide natural variation. Currents must then have swept them from their homeland into their various new abodes, where quirks of a few lucky colonisers then became adapted to the local conditions—left-handed, right-handed or uncoiling. These quirks of colonization are of the kind that biologists like to call 'the Founder Effect'. These little founders had left behind a big effect on their descendents.

A pineapple pin-up

If *Nomgoliella* looks a bit like a garden snail, then the next fossil to turn up in our samples somewhat resembled a modern limpet. Both snails and limpets are members of the same group of snaily molluscs, called the gastropods—both have a slug-like creeping foot and a single shell for protection sat on the back of the body. And both have a body mass that has been twisted round, rather like the torso of the classical Greek statue 'the Discus thrower'. But there is no evidence for such 'torsion' in the torso of *Nomgoliella* nor in its close relatives. So we must incline more closely towards the following view—*that not all that aspires to a spiral is a snail.* That adage, if we accept it, means we must start to wonder whether all those coiled up shells from the earliest Cambrian were really those of proper snails at all. If they were not, then something rather twisted seems to have been going on.

This limpet-like fossil from Mongolia is called *Purella* (see Figure 9). It can be found around much of the world in

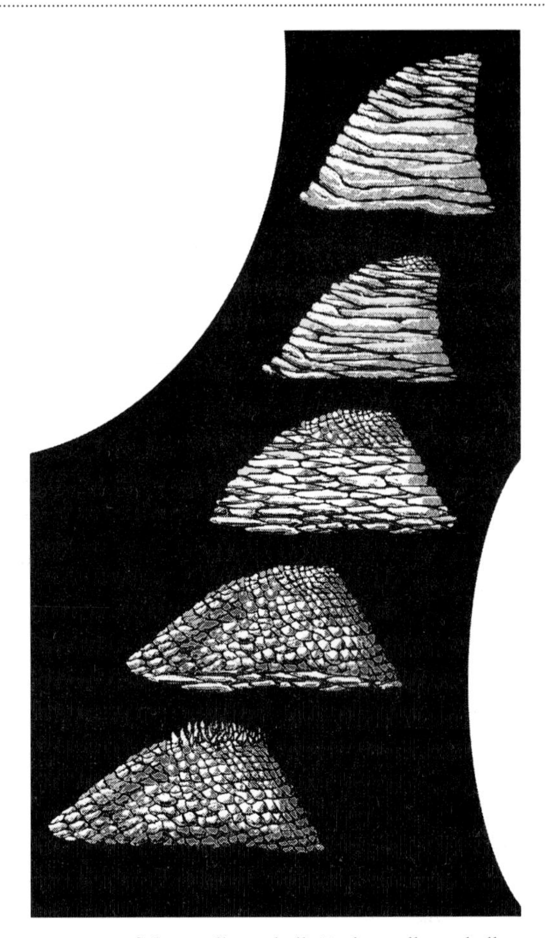

Figure 9. Emergence of the mollusc shell. Early mollusc shells can be reconstructed as part of an ascending series, as drawn here by the author, starting with ancestral *Maikhanella* at the bottom, with a shell of numerous hollow, chalky spines. Gradually, the spines become more elongated and fused, through the forms known variously as *Canopoconus* and *Purella*. At top is shown cap-shaped *Latouchella*, in which the spines have become totally fused into a single shell. These little fossils are usually about 1–3 mm in diameter. Fossils like these have provided some of the earliest evidence for the Phylum Mollusca.

rocks above the first *Anabarites* tubes and below the first brachiopods. Although *Purella* seems to have been a more ancient fossil than snaily *Nomgoliella*, it is easy to see some characteristics that they share. Both have a single, lop-sided shell, shaped like a little pixie hat with its top-knot curled over. But the pixie hat of *Purella* expands rapidly and has only a little curl-over on its top-knot, rather like the cap of Liberty on a French coin, while the pixie hat of *Nomgoliella* is rather narrow and sharply coiled round, more like a French horn. Now, it is not so very difficult to get from one of these shapes to the other. That is because there is a splendid formula—decoded by Dr David Raup of Chicago—that shows how, by means of a slight tweak to three main parameters, one can transform—that is to say, grow, or even evolve—the architecture of almost any kind of mollusc shell we fancy on the computer screen.[81]

Interestingly, many of the intermediate forms that we would expect to find between *Purella* and *Nomgoliella* can indeed be found preserved in rocks of this age in Mongolia, almost exactly as Charles Darwin would have hoped. This lineage also has some delightfully euphonious names, too, which can be arranged in roughly the following sequence: from *Purella* to *Bemella* to *Latouchella* to *Archaeospira* and thence to the younger forms of *Nomgoliella* and *Aldanella*.

But it is the spectrum of forms leading backwards in time from ancient *Purella* that really catches the imagination here. What we find is that this limpet-like fossil grades almost imperceptibly into a form that looks much more like a spiky pineapple. These spines also give it something of the appearance of a modern *Chiton* shell, which has a skirt of spines around its margins. Intriguingly, Charles Darwin had mused upon the scarcity of fossil *Chiton* shells in the rock record. That puzzled him because their seeming absence from the rocks was in such marked contrast to what he had seen of their abundance along tropical shorelines today.[82] But here at the base of the Cambrian, we do indeed meet a rather *Chiton*-like fossil. It is called *Maikhanella* after the Maikhan Uul gorge nearby where it was first found. Microscope work shows that the cap-shaped shell of *Maikhanella* was actually constructed from the fusion of dozens of hollow, chalky spines (see Figure 9). When each of these spines is examined in detail, it can look something like an African vegetable called an okra. Like an okra fruit, each spine was long and hollow, and provided with flattish facets down its flanks. These facets seemingly allowed each spine to nestle snugly against its neighbours. Interestingly, such spines have also been found in isolation in the lowest Cambrian rocks of Mongolia, where they have received a series of almost unpronounceable names, such as *Siphogonuchites triangularis* and *Paracarinachites parabolus*. It seems that the smaller a fossil gets, the bigger its name must become.

In Mongolia therefore, as in China, the earliest molluscan shells seem to have consisted of either isolated or amalgamated stubby spines, as in the manner of *Maikhanella*, or of more plate-like spines, as in a possible intermediate form called *Canopoconus*. And sure enough, the flanks of limpet-like *Purella* also show traces of such an armoury, looking like the fused plates on the back of one of Darwin's famous tortoises. In other words, we can arguably see, in these forms, a sequence of types which could have led towards the evolution of the single solid shell of snaily old *Nomgoliella*.[83] If we take this to be so, then the evolutionary sequence in mollusc shells may have run along a path something like this:

The spines	
Of early molluscs	
Were rather loosely packed.	*Siphogonuchites*
But spines with mineral armour	
Did fewer predators attract.	*Maikhanella*
So those spines packed close together	
Till they were tightly clinkered flaps.	*Canopoconus*
Those spines then fused—forever	
Sheltering molluscs 'neath 'pixie caps'.	*Purella*
Tighter coiling led to 'rams horns' made from this coat of mail	*Latouchella*
But 'tis important to remember—not all that aspires to a spiral is a snail.	*Nomgoliella*

An Element of Hope

As we have seen, Charles Darwin was greatly vexed by the abruptness with which the animal phyla, such as brachiopods and arthropods, seemingly appeared in the rocks near to the start of the Cambrian period. This is how he actually put the puzzle firmly before our eyes:

> The abrupt manner in which whole groups of species suddenly appear in certain formations, has been urged by several palaeontologists, for instance, by Agassiz, Pictet, and by none more forcibly than by Professor Sedgwick, as a fatal objection to the transmutation of species. If numerous species, belonging to the same genera or families have really started life all at once, the fact would be fatal to the theory of descent with slow modification through natural selection. For the development of a group of forms, all of which have descended from some one progenitor, must have been an extremely slow process; and the progenitors must have lived long ages before their modified descendents. But we continually over-rate the perfection of the geological record, and falsely infer, because certain genera or families have not been found beneath a certain stage, that they did not exist before that stage. We continually forget how large

> the world is, compared with the area over which our geological formations have been carefully examined; we forget that groups of species may elsewhere have existed and have slowly multiplied before they invaded the ancient archipelagos of Europe and of the United States. We do not make due allowance for the enormous intervals of time, which have probably elapsed between our consecutive formations—longer perhaps in some cases than the time required for the accumulation of each formation. These intervals will have given time for the multiplication of species from some one or few parent-forms; and in the succeeding formation, such species will appear as if suddenly created.
>
> I may here recall a remark formerly made, namely that it might require a long succession of ages to adapt an organism to some new and peculiar line of life, for instance to fly through the air; but that when this had been effected, and a few species had thus acquired a great advantage over other organisms, a comparatively short time would be necessary to produce many divergent forms, which would be able to spread rapidly and widely throughout the world.[84]

In doing so, Darwin was drawing attention to three curious axioms, which we may attempt to list as follows:

1. Fossil animals appear abruptly.
2. The geological record is full of big gaps.
3. Big steps in evolution must take place very slowly over a long period of time. So this could be taken to mean that the fossil record of the Cambrian explosion must be very incomplete.

Many biologists seem to have taken these three points and the conclusion completely to heart and, in so doing, dismissed the story of the fossil record. But those who do that must deeply misunderstand what the fossil record is trying to tell us. They propagate a slur not only against the early fossil record but also against the nature of evolution itself.

By the time of our own work, Stephen Jay Gould had already demonstrated that axiom number three—big steps take place very slowly—was both conceptually and geologically false. A major aim of our work in Siberia, China and Mongolia was to test these assumptions—and especially the conclusion that the fossil record must be very incomplete—by using the latest analytical techniques.

Intriguingly, much of our work in Mongolia was actually made possible by discoveries some two centuries earlier by Darwin's own chemistry teacher, Professor Thomas Charles Hope of Edinburgh. From Darwin, we learn that Hope's lectures were some of the few in the natural sciences he really enjoyed. Even so, they seem to have had rather a modest effect upon him—he seems to have avoided chemistry through most of his later research. But Professor Hope deserves our attention for another seminal contribution to our story—he discovered that important element called strontium.

Hope first stumbled upon strontium in 1787, when analysing a mineral called strontianite that he had recently collected from near the town of Strontian in the western Highlands of Scotland. During my childhood, this element was often in the news because of its radioactive isotopes, which were turning up in our school milk in the wake of atomic tests by both East and West. That early warning signal about radioactive milk gives a bit of a clue as to why strontium was able to help us in Mongolia—strontium has both stable and radioactive isotopes. Happily, the ocean's reservoir of strontium isotopes is very large indeed and it changes very slowly with time. But change it does—progressively over tens of millions of years. At some time in the distant past, when new ocean basins were growing we find, from the rocks they laid down, that ocean waters and their sediments became quite markedly enriched in strontium-86. That is the condition we find for the early Earth, when the planet was still 'incontinent'—the continents were both small and young. At later times, though, when the continents had become both larger and more active, then newly emerging land masses were able to deliver more and more of the heavier isotope, strontium-87, into the ocean basins. More, in fact, than the oceans were able to produce of strontium-86. Geologists have found this shift in the balance between the isotopes 86 and 87 to be very useful indeed. That is because they can be used to reconstruct a kind of 'contest' between the influence of growing ocean basins on the one hand and of rising mountains on the other. Most famously, this ratio has been used to document the birth of the Himalayas, which were pushed up as a result of a collision between India and Asia. In this way, the Cambrian and Ordovician seafloors and their fossils now sit as much as eight kilometres high in the sky, as my colleagues Mike Searle and Owen Green have so elegantly shown. The huge Himalayan mountain range, and the monsoonal climate that it helped to generate, was to bring about the shedding of vast amounts of strontium-87 into the ocean basins, thereby changing the global signal of seawater, and even of the seaside surf in which you swim.

But we can also use strontium isotopes to estimate the age of a rock, by making comparisons between our data and standard curves obtained from the rock record around the world. Something rather similar can be done with the isotopes of carbon in rocks—such as chalk, or shell or limestone—as we shall later see. Our plan, therefore, was to extend such chemical tests as far back in time as we could—to help us build up a strontium curve across the great Cambrian explosion. And that is where the clever bit comes in. That strontium curve could help us to test whether large amounts of time really were missing—as Darwin had implied—in rocks found just below the earliest animal fossils. Such a test would also provide us with signals that could correlate the earliest animal fossils in different parts of the world. Strontium, then, provided us with the Element of Hope.

As we have seen in Siberia, a great evolutionary leap seemingly occurred near the base of the Tommotian stage. But Darwin had warned us against such phenomena because they may be brought about by big gaps in the geological record. That could mean the 'Tommotian leap' in the Cambrian explosion was simply the product of a prolonged gap in sedimentation during a gradual evolutionary radiation. Worryingly, the cliffs at both Ulakhan-Sulugur and Dvortsy did show us evidence for swallow-holes on the ancient seafloor at the start of the Tommotian. That could be taken as evidence for the exposure of the seafloor, to form an ancient land surface. And that, in turn, could mean that millions of years were missing, through erosion, from the story in Siberia.

Testing the truth of this 'Tommotian leap' in the Siberian fossil record was therefore the object of our first quest, making use of the elements of strontium and carbon. If there had been a big gap there, we might expect to see a big jump in the strontium isotope curve. So we sent the material off to Lou Derry in America for analysis. And that is exactly what Lou found. There was a rapid jump in the strontium isotope record at the very base of the Tommotian stage. It lay precisely at the level where the first brachiopods and archaeocyathan sponges had appeared. Clearly, then, some of the record was likely missing from the story of life in eastern Siberia. But when we then sent off the Mongolian material to Graham Shields in Switzerland for comparison, he found a very different story. He found that the stories of strontium in the rocks seemed to mirror the rate at which species appeared in the fossil record in each region. Going down the cliffs in each of the regions, the story seemed to us to be like this:

In the rocks of Siberia and also in China
The successions seem relatively thin
And there is a very rapid explosion
In the fossilized remains of molluscs.
As well as this, a rapid jump is seen
In the strontium and carbon isotopes
At the level of the Tommotian stage.
Beneath this is a
Gap, and then
Fewer fossils,
No Molluscs,
Or—nothing.

In the rocks of Mongolia the succession is relatively thick
And the rate of deposition was much more gradual
As can be seen from the strontium isotopes
And also in the carbon stable isotopes.
So that there is a more gradual
Increase in fossilized mollusc
Remains, suggesting that
The fossil record here
Is more complete
And that the
Cambrian
Explosion
Of shelly
Fossils
Was
Real
Here.

Modern uranium-series dating shows that all this evolution—from spiny molluscs at the start to snaily molluscs, brachiopods, and trilobites near the top of the hills, spanned about 15 million years. That is a very short period of time indeed in geological terms—given that the planet is some 4,560 million years old and the universe is about 13,700 million years old. But the Cambrian explosion was clearly not an instantaneous event. As Darwin himself had predicted, there should be, and there was, a gradual increase in the number of different species within the invertebrate animal phyla at about this time:

> gradual increase in number of the species of a group is strictly conformable with my theory; as the species of the same genus, and the genera of the same family, can increase only slowly and progressively; for the process of modification and the

> production of a number of allied forms must be slow and gradual,—one species giving rise first to two or three varieties, these being slowly converted into species, which in turn produce by equally slow steps other species, and so on, like the branching of a great tree from single stem, till the group becomes large.[85]

But it all depends on what we mean by 'gradual'. To a biologist, 15 million years sounds like a gradual 'ffffffffffffffUT!' But to a geologist used to working in Deep Time, it can sound like an explosive 'BANG!'

Why, though, did all this happen at all? Could it have been triggered by a single intrinsic event—like the evolution of eyes, as Andrew Parker has suggested?[86] The problem here is that the emergence of features such as eyes—like those we have seen in *Fallotaspis*, and mimicked in *Microdictyon*—seems to lie quite high and late within the cascade of events. And even the humble jellyfish can have vision. Maybe we should be looking for some more basic evolutionary trigger, such as the evolution of the tubular gut with its mouth and anus. Putting aside, for a while, the mischievous thought that it was *the evolution of the anus* that caused the Cambrian explosion, it is time to take a closer look, like a demon dentist, into the jaws of a primitive predator.

A terror emerges

Eager to see the Mongolian fossil remains for myself, I peered curiously down the barrel of the microscope. There, as expected, was a tube with a clover-leaf cross section like the one we met in China—called *Anabarites*. Wherever this fossil turns up, we seem always to be very close to—or even right at—the very start of the Cambrian explosion of animal skeletal fossils. Curious. Why did our old *Anabarites* and (the younger) *Nomgoliella* need mineral shells at all at this time? Many beasts get by without shells or skeletons today, as we can see whenever we walk down the garden path. There, we can spot naked slugs and earthworms, centipedes and woodlice, all getting along without any mineralized skeletons at all. The latter are living alongside birds, snails—and ourselves—all provided with mineral skeletons. Put another way, what are skeletons and shells really *for*? And can the answers to that question tell us something about the *cause* of the Cambrian explosion? In search of one possible answer, we may turn from *Anabarites* towards a second kind of fossil that can be found in Mongolia. If you're at all squeamish, look away now.

After searching through tray after tray of harmless little *Anabarites* down the microscope, the occasional occurrence of *Protohertzina* can prove a bit of a shock. This fossil is most decidedly *not* a dwelling tube. It looks more like the tooth of a tiny fossil *Tyrannosaurus rex*. Curved and scimitar-like with a wicked little point at its tip, these tooth-like 'protoconodonts' are sometimes found preserved together in clusters, though they are mostly encountered alone.

The owner of these sinister scimitars may well have been something like the living Arrow Worm, whose sickle-shaped blades are arranged around its mouth in clusters, in the way that a portcullis might be arranged around a castle gate (see Figure 10). These torpedo-shaped worms are famous today both for their abundance and their rather nightmarish predatory habits. They will swim through the water column in search of suitable prey. Being as transparent as glass, they can arrive unseen, to tranquilize their prey by injection of a nerve poison, called *tetrodotoxin*. The poor victim then collapses into a torpor, allowing the Arrow Worm to sink in its teeth and carve up lunch.

Equally interesting is the evolutionary position of Arrow Worms in the tree of life. They may be among the most primitive

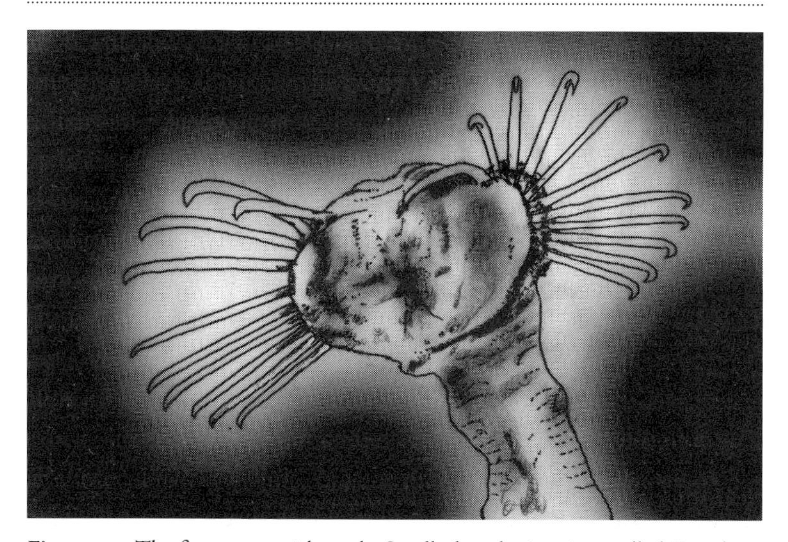

Figure 10. The first terror with teeth. Small phosphatic spines called *Protohertzina* are often among the first skeletal fossils to appear in rocks around the world, at the start of the Cambrian, some 542 million years ago. These little fossils are usually about 1–5 mm in length. They are here reconstructed by the author as part of the feeding apparatus for some kind of predatory worm, perhaps allied to the modern Arrow Worms in the Phylum Chaetognatha.

of any animals with bilateral symmetry alive today. As we have seen, sponges need have no symmetry at all, while corals and jellyfish typically have radial symmetry. Most other animals, like ourselves, have bilateral symmetry—meaning that there is only one mirror plane of reflection. Arrow Worms have bilateral symmetry, which implies that they are more advanced than corals. But they seem to have missed out on some useful bilaterian inventions. For example, they have no blood system for circulation, no gill system for respiration and no excretory system for the removal of waste. The unwanted remains of an Arrow Worm's lunch are, therefore, rather disconcertingly, excreted through its skin.[87]

So why does *Protohertzina* matter so much? The first answer to this question is that *Protohertzina* provides the first clear signs we have for the emergence of predation in the fossil record. Predators—like arrow worms and aardvarks—need other animals to prey on, such as antelopes and ants. And these other animals are usually themselves feeding on plants. This means that *Protohertzina* provides the first physical symptom of ecological tiering in the fossil record—of animals feeding on animals that fed upon plants that fed upon air and sunlight. The ecosystem was seemingly starting to self-organize into some kind of food chain.

But the second answer is equally fascinating: *Protohertzina* first appears at the very start of the Cambrian explosion, alongside *Anabarites* and *Maikhanella*. We know this because we have traced it from Mongolia, through China and into India and Iran. It also turns up in Siberia and Canada at this low level. That means that some of the earliest skeletons in the fossil record turn up at about the same time as the earliest predatorial jaws. It is like finding the disturbing remains of an arms race—swords with shields, guns with tanks, bombs with bomb shelters—in an archaeological dig.

The evolution of a mouth and jaws might be seen as a single and inevitable step in a war of ecological escalation, of course, rather than the very start of the explosion. To gain a better picture of the dynamics involved, it is helpful to travel away from Central Asia and southwards towards the Himalayan mountain belt. But not just to look at fossils.

To look at landslides.

The Cambrian Cascade

Along the hillsides around Kunming in south China, the peasants we met back in 1986 were hungry for fuel. Once, when it came on to rain, we met up with a breadwinner struggling with a huge pile of logs on his back, to feed the family fire. He stopped to stare at me, clearly amazed at my western appearance, for a full sixty seconds. I looked around, a little embarrassed, and then began to notice that many of the trees had disappeared from the valley slopes around us. Glancing again at his logs, it was easy to guess that he was helping to deforest the hillside. Without trees, the soil was in revolt; it was sliding down the hillsides, often blocking traffic, on its long journey back to the sea. On one such journey, starting near Kunming, we found that the resultant landslide had caused three overturned lorries and a crashed motorcycle. The rivers were swollen with red soil being wasted away in this fashion.

Vast areas of Asian hillside are beginning to slide down to the sea in this way. Curiously, these sliding beds—with their Cambrian fossils—can be traced around the mountains from Yunnan and Sichuan, through Vietnam (where they are greatly crushed and altered) and into to the Himalayan foothills of India, Kashmir, and Pakistan. One of the most famous sections for the study of the Cambrian fossils lies at the top of a long and winding road north of Dehra Dun in Uttar Pradesh in India, where I was invited by Dhiraj Banerjee of Delhi in 1990. Here, at the top, in the cool fresh air above the great plain, lies the little hill town of Mussoorie. It was in places like this that families of the British Raj once liked to keep cool during the great heat and humidity of the Monsoon season. And here can be seen much the same kind of sequence of events that we have encountered at Wangjiawan in China, some thousands of kilometres to the east: phosphate rock with *Anabarites* and *Protohertzina*, passing down into limestones with algal remains, and then down into glacial deposits and sandstones, all many thousands of metres thick.

But it is the mathematics of the Mussoorie road—or rather the hillside that it cuts—that we have come to contemplate here. There is no better place to picture the dynamic and restless nature of our planet. The Himalayas are steep and high because the mountains have risen up faster than the rate at which rivers can carry their rocky detritus back down to the ocean. This means that the slopes of the Himalayan foothills can be very steep indeed, so steep and unstable that rocks and soil do not long remain at rest—they are always close to their tipping point.

Peering over the steep side of the Mussoorie road gives a clue to the processes going on here. The flanks are covered with fans of debris, some small and some very large. Every now and then, one of these fans slumps down on to the road below, or cuts it away from beneath, bringing traffic to a halt for anything from hours to days. Not surprisingly, the engineers have made a note of the frequency and magnitude of these landslides. From this it has become clear that there are large numbers of little landslides—about the size of a shovel full—ranging up to large but rare ones that are up to 10,000,000 cubic metres in volume.[88] Theorists have found that such landslides will propagate wherever a hill slope reaches a supercritical state. Something like this can be simulated in the kitchen, by tilting a sugar bowl and then watching how the sugar moves when at a critical angle, usually all at once and across the kitchen floor.

It is in the foothills of the Himalayas that we can try to bring a rather wider range of phenomena—shells and teeth, predators and prey, phosphate and landslides—into some kind of perspective—or at least an attempt at one. Each of these has some bearing on that great mystery we call the Cambrian explosion. And each of them is in some way connected. And connectedness lies at the heart of the matter. Let me try to explain.

As we have seen, the fossil record suggests that the emergence of the major animal phyla—from chordates to worms—could have taken place very fast indeed, in geological terms, say between about 545 and 530 million years ago. That is much less time that Darwin was expecting. But there may be good grounds for believing that complex systems can self-organize very rapidly indeed, and landslides of the Himalayas provide us with a crude kind of metaphor for how this might have been possible.

In the manner of Admiral Nelson discussing his impending battle plan of Trafalgar, using a pile of beach sand, plus salt cellars and spoons, I like to visualize it as something like this. The vast mass of the Himalayan mountains—formed by this pile of sand over here—stands for the biomass of primitive multicellular animal life, building up during the prelude to the Cambrian. This build-up drove the biosphere—the sand pile—further and further away from equilibrium. As an aside, we may speculate that the carbon cycle became unstable. Or the climate became too extreme. Anyway, it was something to do with *connections within the system itself.*

At a certain point, we notice that the bulge 'goes critical' and the pile collapses, calving into a series of avalanches—landslides—that subdivide and isolate into a series of new masses. These landslides stand for those species that were to become the ancestors of the new animal phyla. But the calving of new phyla does not return the biosphere towards equilibrium, in this model. Cascades of other avalanches continue to take place, some of them quite large—the animal classes and orders—but most of them very small—the animal genera and species. These processes are still ongoing.

Remember that we are trying to contemplate why the evolution of the animal phyla seemingly took place *almost all at once,* changing a seemingly monolithic biosphere towards one comprising a burgeoning number of new animal species. Seen from this perspective of an avalanche, each speciation event may be compared with a landslide, slicing off slivers from a relatively monolithic ancestral mass of animal ancestors to form an array of new entities and domains. Each avalanche is driven towards one of many basins of attraction, like river valleys, including those large attractors we now call the animal phyla. Seen in this way, this Cambrian Cascade conjectures that the explosion of animal phyla was the inevitable but rapid culmination of an earlier step—the evolution of multicellular organisms themselves. Like the lining up of grains in a landslide, its precise timing was likely due to more than a single factor. A single, external driving force—what we might call a deterministic explanation, such as seawater chemistry or oxygen—seems unlikely to provide the final answer. The behaviour of complex systems means that the finger points to a chance interlinking of several factors *within the complex system itself*.[89]

It might be complained that this metaphor—of the landslide—is inappropriate and more relevant to ecosystem collapse and the death of the dinosaurs than it is to times of rapid speciation and to ecosystem construction. But the best analogies must typically come from outside the system if we are to avoid the dangers of weak inductive logic and circular reasoning.[90] And use of the landslide metaphor can change with perspective—one creature's mass extinction will be another creature's window of opportunity. Take the collapse of the Roman Empire as a case in point here. Imagine that your ancestors were tax collectors—and some of them probably were. The collapse of empire after AD 409—into lots of tiny autonomous kingdoms—could have been seen as a bad thing from their point of view. But now change the perspective, to that of a small town mercenary soldier. For such a person the empire's demise must have seemed full of exciting new potential, pointing the way towards self-government and small kingdoms needing lots of freelance soldiers. We tend to think of Dark Ages as a 'bad thing'. But it all depends on perspective.

In search of an *Übermonster*

Many natural phenomena behave in the manner of landslides, as the Danish mathematician Per Bak and others have shown. Landslides, earthquakes, traffic jams, and stock market crashes are just a few of the examples that have been studied so far. In each of these types of system, it is largely impossible to predict exactly when a significant event will take place, and how large it will be, even though the probability of such an event is high. That is because events such as avalanches are propagated by sensitivity to initial conditions—a little less friction here, a little too much liquidity there, and so on.

The mathematics of complexity is consistent with the idea that the Precambrian world may have been brought to an end by something like an avalanche within evolution itself. Many of the features that we have seen at the start of the Cambrian could have been triggered by small happenings, and most importantly, by happenings that were *intrinsic to the system*. Human history and warfare is full of such contingencies—the unplanned visit by King Harold to William of Normandy in 1065; the chance meeting of Archduke Franz Ferdinand with a gunman on a street corner in 1914; the chance survival of Adolf Hitler during the Great War. When Prime Minister Harold Macmillan was asked what he feared most in politics, he famously answered: 'Events, dear boy, events.' Out of such happenstance come forth more familiar and agreeable things, too, such as democracy and Darwin, jazz and the jumbo jet. In other words, big consequences do not need big things to trigger them. Even little things can have big consequences. That is an alarming thought. But it seems to be true.

An avalanche model for the Cambrian explosion homes in upon driving forces that emerged *from within the biological system itself.* It predicts that the story will turn out to resemble a series of feedbacks that could have run something along these lines:

Variation
P l u s
Selection
Led to a
Mouth and gut.
The introduction
Of an anus meant that
Food, through gut, was put,
Some humbly swallowed mudcakes.
Some gobbled sea weed in.
A few rapacious carnivores.
Were met with chewy skin.
Predators raised the stakes, therefore, with jaws of toothy bite.
Prey then fought back with mineralized shells of fluoroxyapatite.
Predators toughened their teeth yet more with carbonate of lime.
So creatures burrowed down below to avoid the threatening luncheon time. . . . [91]

In terms of our analogy of the card game, the Cambrian Cascade is more like a House of Cards than a proper card game. It says that a series of events, seemingly quite small ones, happened during the stacking of the cards which led to a kind of avalanche—perhaps the greatest avalanche ever seen in evolution. And that cascade was so large that it not only raised the visibility of certain players, it also changed the rules of the game. If there is some truth in this hunch, we should be able to find in the earlier fossil record, some evidence for the 'foreshocks' themselves.

CHAPTER FIVE

A WORM THAT CHANGED THE WORLD

Eye Spy

In May 1983, it fell to me to drive a ramshackle minibus containing four Russian and four Chinese scientists across England and Wales. We had just completed our 'Country House murder mystery meeting' at Burwalls. But the mood in the bus was dark. We had voted on the placement of the Precambrian–Cambrian boundary. And the Chinese delegates had almost overthrown the Russian case (and in the following year, they actually succeeded in doing just that). Clearly, they meant business. Added to that was the fact that two of our passengers—Boris Sokolov and Xing Yusheng—were of impressively senior rank. Had I accidentally 'rolled' the minibus during the week-long trip, some said, I could have triggered World War Three.

Seated at the front beside me was the large and amiable Russian interpreter. He clearly sensed my concern, and sought to reassure me, in a deep and resonant basso-profundo voice: 'I will help you. Let me be your Map Reader!' While I was digesting this unexpectedly kind offer, he swabbed the outside of the front windscreen with a hand like a bear's paw—not only on his side of the vehicle but also on mine, all without leaving his seat. The windscreen thus cleansed of blemishes, he beamed with satisfaction and snapped open a black attaché case resting on the seat beside us. A solid phalanx of photographic film was arranged in neat rows. Then out came a shiny silver Russian camera with a flourish. He looked into my face quizically, his eyes twinkling.

It is true that our Russian interpreter could have been a keen naturalist who merely planned to photograph the darling buds of May. But as soon as we drove across the Severn Bridge, my 'Map Reader' started to click away with impressive determination, pointing the nozzle of his camera hither and thither. Not towards the breezy scene around us, of course, but directly at the bridge superstructure high above our heads. An hour later and it was the masts of hilltop radio receivers that received similar affectionate attention. By the end of the trip he had also snapped a nuclear power station, a military airfield, and me.

Strangely, this camera would come out of its resting place just a little before each object of interest hoved into view. True, we had Ordnance Survey maps for us to look at, but these could not have provided the clues because they rarely show radio masts and airfields. Puzzled, I devised what I thought was a cunning manoeuvre. I would flick the front windscreen wipers into furious action every time the camera came out. His index finger, and my windscreen-wiper did battle for days as we crossed a patchwork quilt of green, yellow and brown fields. This rather curious behaviour—on both our parts—may, perhaps, be excused because it took place at the very height of the Cold War.[92]

Tweedledum and Tweedledee

Like ancient hunters, our plan on this trip was to sniff out the ground for animal tracks—the spoor of the earliest animals on the planet. Tracks can yield truly first rate clues to the decoding of Darwin's Lost World. And that strange truth first emerged in the nineteenth century, as the result of a geological prize fight between two geo-giants—Professor Adam Sedgwick of Cambridge and Sir Roderick Impey Murchison of the UK Geological Survey. They were the Tweedledum and Tweedledee of Victorian geology: 'Contrariwise, if it was so it might be; and if it were so, it would be; but as it isn't, it ain't. That's logic.'[93]

Murchison, a veteran of the Napoleonic Wars, was much given to chasing hounds across the Welsh Borderlands describing, almost at the gallop, some of the world's oldest known fossiliferous rocks.[94] Named by him after the old Celtic kingdom of Siluria, where Caratacus had once led resistance to the Roman invaders for nearly ten years, he liked to think of *his* 'Silurian' rocks as containing the origins of life itself. Sedgwick, on the other hand was a twinkly old Cambridge don, a clergyman who didn't seem to be much good at finding fossils, or even to care for fossils very much. He believed that *his* old rocks, which he was tracing across much of North Wales, had been laid down before the creation of life, meaning they were therefore lacking in fossils. These rocks he called both 'azoic' from their lack of life, and of course 'Cambrian' after the Roman name for Wales.

That was all fine and dandy until it was shown by the Frenchman Joseph Barrande that Sedgwick's Cambrian rocks actually contained fossils such as trilobites, not only in Bohemia (in 1846) but also in Wales (in 1851).[95] Near Dublin, Edward Forbes then described some beautiful but strange markings called *Oldhamia*, looking like rows of tiny crow's feet. In 1846, he thought of them as the remains of some kind of soft coral. Even Darwin made some oblique reference to this emerging work in the *Origin of Species*:

> To the question why we do not find records of these vast primordial periods, I can give no satisfactory answer. Several of the most eminent geologists, with Sir R. Murchison at their head, are convinced that we see in the organic remains of the lowest Silurian stratum the dawn of life on this planet. Other highly competent judges, as Lyell and the late E. Forbes, dispute this conclusion. We should not forget that only a small portion of the world is known with accuracy. M. Barrande has lately added another lower stage to the Silurian system, abounding with new and peculiar species.[96]

To which was added by 1872:

> and now, still lower down in the Lower Cambrian formation, Mr. Hicks has found in South Wales beds rich in trilobites and containing various fossil molluscs and annelids.[97]

Murchison was hugely un-amused by all these unwonted developments. He moved with the panache of a cavalry officer, outflanking Sedgwick to progressively swallow up the latter's ancient Cambrian system. The origin of life was to be Murchison's, and Murchison's alone. No wonder he was called the 'King of Siluria'. For a decade or more, the poor old Cambrian slumbered somewhere deep inside the belly of the Silurian, ignobly removed from maps of the Geological Survey. Muffled complaints could still be heard in Cambridge but barely at all in London or beyond. All this explains Darwin's use of the term 'Silurian' in 1859, for rocks we would now call 'Cambrian'. The 'King of

Siluria', victorious, was thereby able to rule over a handsome domain stretching from Much Wenlock in the east to Harlech on the Welsh coast. The new Director of the Survey thoroughly approved of this turn of affairs, not least, perhaps, because he was, by a curious coincidence, a certain Sir Roderick Impey Murchison.

The middle management of geology were, however, not a little aghast at Sir Roderick's presumption. Charles Lyell and John Phillips, both based at colleges in London, began to cast around for a solution. Salvation seemingly came in the form of a palaeontologist called John William Salter. Untutored but remarkably gifted, Salter had two great assets of the kind much needed in this field—a knack for finding great things, and an artistic flair. Keen to test the existence of a truly Cambrian biota, Salter set off—with geological hammer in hand—to begin a three-week search, from Wenlock to Wentnor in the Welsh Borderlands. Revelation took place within a deep valley in the Longmynd hills of Shropshire. These rocks had been notionally accepted as both azoic (without life) and 'Cambrian' by Sedgwick as well as by Murchison. But it was here that Salter found what he took to be the remains of wormy activity on the seafloor as they went about their daily business. These were pairs of pits which reminded him (wrongly as it happened) of the openings of U-tubes like those made by the modern lugworm *Arenicola*—so he called them *Arenicolites*. He also found what he mistakenly took to be a trilobite which he called *Palaeopyge*. 'The Cambrians are not barren of organic life' he excitedly wrote to Murchison. But what Salter had actually stumbled upon was not worm tracks at all. Nor were they even Cambrian. *They were the first genuine Precambrian fossils ever to be named and described.* The year was 1855.

Scribbles in the sand

Every game needs *rules*, even the game of Darwin's Lost World. Unfortunately, the great rule book has been missing for the last 540 million years, ever since the start of the Cambrian period. As in any game of cards, though, a good way to guess the nature of the game is to sit down at the green baize table and take part in successive rounds. In this way, steadily watching how the cards fall, we may hope to gain important clues.

Thus far, we have played three rounds in our hunt for Darwin's missing fossils, travelling to Siberia, China, and Outer Mongolia to investigate patterns preserved in the fossil record, in the form of shelly fossils like *Aldanotreta* and *Anabarites*. When prodded with the right questions, these skeletal fossils have shown us that they will happily yield up some valuable clues. But some of the players at the Burwalls gathering in 1983 were unconvinced that small shelly fossils would prove to be our best guide to events at this time. Shelly fossils cannot, for example, be found in large parts of the geological record because of their unfortunate need for special conditions of preservation. While conditions were clearly just right in parts of Russia and China, much of the evidence has faded away to nothing in rock successions in other parts of the world. That was proving to be the case, for example in the slate belt of Wales and in the Rocky Mountains of America. A new line of evidence was therefore going to be needed that was robust enough to be preserved in the rock record and common enough to be traceable around the world. Luckily, such evidence was about to be unearthed along the shorelines of maritime Canada.

This new evidence was to take the shape of markings that can look a bit like 'graffiti', not unlike those we can find scribbled on the walls of a city subway. These markings have not, however, been idly rendered in lurid day-glo paint. They are messages written with great earnestness that now lie sleeping within the pages of the fossil record. Intriguingly, the oldest such markings—the oldest convincing ones, at any rate—date from the very beginning of the Cambrian period (see Plate 8).

John William Salter had made a start on the study of worm tracks back in 1855. But it took a further hundred years to reveal that such markings may hold vital clues towards the decoding of Darwin's dilemma. Hints of this began to emerge again in 1956, when the German palaeontologist Dolf Seilacher began writing about rocks from the Salt Ranges of Pakistan. Seilacher discovered that Cambrian rocks in this area contained traces remarkably like those along a modern shoreline. Plenty of worm trails and arthropod scratch marks were waiting to be discovered, at least in Cambrian and younger rocks. But in older, Precambrian rocks, however, there did not appear to be many such markings at all.

The pioneering work of Dolf Seilacher later blossomed into the discipline called *ichnology*—the study of modern and fossil animal tracks and trails. It would be wrong to call it a new discipline because the interpretation of such markings harks back to our Stone Age ancestors, who often relied upon their skill in reading tracks for their supplies of fresh meat. Reading the runes of footprints is a skill that can still be found today among the few remaining hunter-gathering societies, like the !Kung of Africa and the Aborigines of central Australia.

Using something like those ancient skills, Seilacher was able to observe that the 'animal spoor' change dramatically at the end of the Precambrian period, from patterns that seem rare and strange towards patterns that seem abundant and familiar. One of his followers in England was geologist Peter Crimes. He followed this theme in rocks around the world. At the Burwalls meeting in 1983,

Peter was therefore able to report, with geologist Terry Fletcher, the discovery of a seemingly complete sequence of animal 'footprints' in Newfoundland, exposed along the north-eastern seaboard of Canada.[98]

The rocks from New York to Newfoundland, and from Wales to England are found to have shared a strange story, especially during late Precambrian to Cambrian times. Indeed, it was only two hundred million years or so ago—a mere heartbeat in cosmic terms—that these now distant regions were wrenched apart by gargantuan geological processes. That meant that the story in the rocks could therefore be read in two slightly different editions—what we might call an unexpurgated version and an abridged version. The unexpurgated edition lies along the coast of maritime Canada, within a world of bog and moss. That is the land where Peter Crimes had been working, with help from Mike Anderson. But a strongly edited version of the same story was being found on the European side of the Atlantic, in the valleys of Wales and the road-stone quarries of Warwickshire, where I myself was then researching. Strange as it may seem, beds with the very same fossils in the very same sequence through the rocks were starting to be found on both sides of the Atlantic, separated by three thousand miles of ocean.

A good example of this shared heritage is a Tommotian shelly fossil called *Coleoloides* (see Figure 7). It looks a bit like a tiny Narwhal's horn because it comprises a long tapering tube graced by an external ornament of spiralling ribs that probably helped it to anchor in the mud. We have no idea what kind of creature lived inside the tiny tube. I once liked to think it was a miniature tube worm like those that now live beside the scalding waters of black smokers. Whatever its nature, though, it is clear that *Coleoloides* became one of the first animals to become seriously rock-forming.

That is to say, the shells of *Coleoloides* helped to build up mounds of limestone on the seafloor that once stretched all the way along a continuous shoreline from New York to the middle of England.

A further fossil shared between England, New York, and maritime Canada is the earliest trilobite in this region, called *Callavia*.[99] Curious to report, primitive trilobites like this were seemingly rather wimpish swimmers that seldom dared to travel far out to sea. Happily for *Callavia*, though, the wide Atlantic Ocean simply did not exist as a barrier to migration in Cambrian times. In those days, New York was just down the coast from England and both were part of a great chain of volcanic islands now called Avalonia.[100]

Fortune's edge

During the field trip to the Longmynd and Wales, we could see that these successions could help to provide us with valuable clues to the early history of animal life on Earth. But there was a problem: the story in England and Wales—although rich in historic interest—was largely hidden underneath thick layers of turf and woodland. Many pages of the book were therefore missing inland, and the rocks only rarely touched the coastline, where the best stories might have been read. But at the Burwalls meeting in 1983, we were starting to hear that a very similar succession of events was now being found around coastlines on the opposite side of the Atlantic Ocean, in Newfoundland.[101]

Arguably, there is no better place to examine the Cambrian explosion and the earliest animal tracks than the coast of south-eastern Newfoundland, and most especially around a little fishing settlement called Fortune. The coast around Fortune Head is an open-air museum to the history of life.[102] Bleached bones and shipwrecks litter the coastline along this part of the Burin Peninsula. On a foggy day, the shoreline can smell richly of the sea and its tithe of fish and flotsam. One popular tourist map of the region shows the names and dates of the many hundreds of historic shipwrecks. Lonely pebble beaches on this part of the island can be piled high not only with mounds of kelp but also with spars and planks from ships, some of which have foundered during winter storms and summer fogs. These Newfoundland fogs can seriously dampen the spirits of both seamen and landlubbers, too, including geologists. It is no surprise, then, that the spirits of locals are raised by commerce with the nearby islands of St Pierre and Miquelon—both of which have somehow managed to cheat destiny and retain their status as a joint Territorial Collective of France. These islands now sit as provocatively within Canadian coastal waters as might a pair of trollops at a tea party. From shipwrecks to smuggling, and from fogs to performing worms, this is a strangely deceptive terrain.

As in Siberia and China, I was seeking a story from rocks that lie below the trilobites. But no visit to this region would be complete without starting with the truly spectacular trilobites of nearby Fortune Brook. My research students Richard Callow and Alex Liu heard of this amazing deposit at a local bar, while being 'screeched in'—an ancient rite of passage that Newfoundland fisher folk are pleased to inflict on visitors.[103] Here, on the outskirts of Fortune, a stream flows between banks of slate crammed with trilobites, many the size of Newfoundland lobsters. These monsters lived at about the same time as the Burgess Shale creatures—on the other side of Canada—that is to say, in the middle of the Cambrian period about 505 million years ago. But *Paradoxides* is much larger and more awesome than most of the Burgess Shale animals, being up to 30 cm long and bristling with an armoury of up to 60 knife-like spines. The largest of these defensive weapons was a tail like a pike staff that is thought to have acted as a kind of vaulting pole.

With Mike Anderson as my guide in 1989, we drove past Fortune Brook with its giant trilobites, parked the car at Pie Duck Point, and packed our bags in readiness for the long hike across rock and peat bog. The earliest trilobite here is *Callavia*, whose remains can be found in a bay, called Little Dantzic Cove, cradled between brick red limestone and dark grey shales. The next few kilometres of coast bring up to the surface a series of rather drab and largely unfossiliferous sandstones of the Random Formation, of Tommotian age. But here and there, the monotony is relieved by sheets of rock covered in mica flakes that shine brightly in the sunshine like salvers of polished silver.

Walking down through these dun-coloured sandstones, we eventually reached a succession of small bays, each decorated with alternating bands of pink and green, like a great big block of Neapolitan ice cream. Here we could find no evidence for trilobite fossils, suggesting that we were, at last, down in the 'Pre-Trilobite' part of the Cambrian and getting very close to its base.[104] Crawling about on hands and knees, we could spy 'small shelly fossils' in the greenish muds, much like those in the Nemakit-Daldynian of Siberia and the Meishucunian of China. Some of them were new to us, though, and were to prove particularly useful as clues towards an explanation for Darwin's missing fossils.

All stuck-up

Aldanella is a close relative of the primitive mollusc *Nomgoliella* that we met in Mongolia, except that the Canadian example differs in having a shell that is twisted round in a right-handed way rather than a left-handed way (see Figure 8). As with many other early shells of this age, the shell may well have shone on the seabed with the iridescence of mother-of-pearl. Unfortunately, in the rocks around Fortune, the pearly shell of this little fossil has long ago been replaced by a brassy mineral called pyrite. But this petrification has been achieved with such exquisite fidelity that one still can see the growth lines of the tiny snail shell and imagine its pearly lustre.

Platysolenites is not so pretty to look at as *Aldanella*. It had a tubular shell like a drinking straw. But it actually provides two highly important clues for the decoding of Darwin's Dilemma. As we have seen, tubes were widely favoured by the earliest shelly creatures—such as *Anabarites* and *Coleoloides*—in part because they were simple for novices to construct. The first clue here is that we have found a shell that was built by gluing together bits of miniscule silt grains (see Figure 7). This is a very telling discovery. It shows that Daly's Ploy—that the Cambrian explosion was about the evolution of calcium skeletons in relation to changing seawater chemistry—rather misses the point. *Platysolenites* shows that the Cambrian explosion was not just about the secretion of chalky or glassy minerals at all. It was about the evolution of *all* kinds of skeleton, including those made by gluing bits together. Clearly, the minerals of carbonate and phosphate were just one of several options available at that time. Any material would do, even little sand grains lying around on the seafloor. And that would imply that protection from predation rather than changing seawater chemistry, is likely to have been the greater driving force here.

But equally astonishing is the evidence that the shell of *Platysolenites* was probably not built by a multicellular animal at all.

As Duncan McIlroy of the Memorial University in Newfoundland has shown, it bears all the features of a single-celled protozoan. Similar shells, made of grains of sand and silt and glued together by protozoans with exquisite precision, can still be found in stupendous proportions on the deep sea floor today. That is significant because it shows us that the whole of the biosphere, not just the world of multicellular animals, was caught up in the great revolution of the Cambrian explosion. In other words, we should perhaps be seeking an explanation that involves *co-evolution of the whole biosphere.*

The Circus of Performing Worms

As Mike Anderson and I hiked down the section from one bay to the next and towards Fortune, there was a point at which even these small shelly fossils could no longer be found. In these earliest of all Cambrian rocks, some 542 million years old, all that was left behind for us to find were worm tracks—wiggles and scuffs, twists and turns—each preserved on the flaggy bedding plane of sandstone. Millennia of weathering have helped these ancient wiggle marks to stand out on the surface of the rocks. These wiggle marks are called 'trace fossils' because they mark the traces of ancient animal activity (see Plate 8).

When a palaeontologist looks at the wiggles of ancient worms in rocks like these at Fortune, he or she will try to imagine not only their shape but also their movement across the sea floor. This process of reconstruction is rather like trying to recall a distantly remembered circus act. So, how might it feel to take seats at the ringside of this Cambrian Circus of Performing Worms? It might be helpful to try to imagine such a scene (see Figure 11).

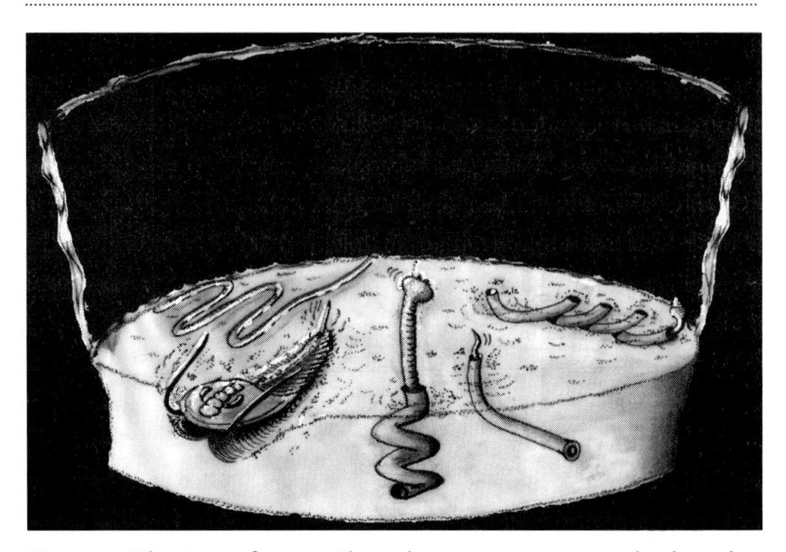

Figure 11. The circus of worms. Shown here are reconstructions by the author of some of the earliest examples of animal activity preserved on the seafloor, after 542 million years ago. These fossils are usually some 3–10 cm in length. From one o'clock and turning clockwise, these fossils include the feather stitch-like feeding trace of *Trichophycus*, the simple tubular burrow of *Planolites*, the spiral trace of *Gyrolithes*, the scratch marks of *Rusophycus* here made by the trilobite *Callavia*, and the meandering feeding trace of *Helminthoida*. The first appearance of such an assemblage in the fossil record is taken to mark the beginning of the Cambrian explosion.

A cheery cacophony announces our opening circus act, during which the 'arena'—our Cambrian seafloor—starts to fill with sand to a depth of several metres. The imaginary ring master invites us, during this arenaceous interlude, to conjure up the amounts of time needed to train the performing animals we are about to witness. We sit in silent appreciation for ten minutes or so, in deference to the four thousand million years that had to pass before the first animals, our ancestors, were to appear upon the Earth.

At last, a spot of light illuminates a patch of sand near the centre of the arena floor. A large worm-like creature attempts to propel itself across the sand. It tries several manoeuvres: first it tries rolling sideways, to leave markings like those called *Palaeopascichnus*. Then it tries stretching and contracting along the length of its body, like an inchworm, to leave behind a long thin trail, like the one called *Planolites* (see Figure 11). For an encore, our wormy artiste disappears rapidly from view, burrowing down below the surface of the sand in the arena. There follows a murmur of expectation—vertical burrowing means that the Cambrian period itself is about to start . . .

A long loud blast—from a conch shell, of course—announces Day One of the Cambrian period. A luridly coloured serpent emerges from the wings and starts to writhe about on the arena floor. For every two lengths it moves forward, the serpent comes back one length and then tacks off again, slightly to one side. This balletic performance goes on for several agonizing minutes so that a trail is left behind in the sand that looks a bit like a series of large blanket stitches (see Figure 11). Here before us, then, is a reconstruction of the worm track taken to characterize the start of the Cambrian period, called *Trichophycus pedum.*[105]

All this whirling and weaving not only stirs up the sand, it also stirs up the crowd. But the audience is advised not to throw in its pennies just yet, for there are stranger things to come. Gradually, as the lights in the arena grow dim again, all we can see is a patch of sand, carefully raked smooth and flat. The sound of a low drum-roll reverberates in the distance. Then, with a crashing of cymbals, a slender black worm rises up from the depths of the sand, making a corkscrew motion with her body (see Figure 11). Nothing like this belly dance has ever been seen on Earth before. The audience breaks out into a round of delighted applause, for this is the maker of the trace fossil *Gyrolithes*—meaning *the twisting rock-builder*.

Next comes a slight change of pace. A trail is left across the freshly raked sand by a serpent that weaves, ever-so-slowly, across the surface of the arena floor, from left to right (see Figure 11). This series of hairpin bends is of the kind that palaeontologists like to call *Helminthoida*. At first, the reason for this almost hypnotic motion escapes us. But during the following encore, we notice that colourful little parcels of food have been scattered, in small patches, across the arena floor. By this weaving action, the worm-like creature that made the *Helminthoida* trail was able to gather up these parcels and reap a richer reward from patchiness than a straight line across the arena could ever have done.

But the final act strikes a more mordant tone. The chattering of the crowd in the arena drops to a whisper as a school of lithe and balletic performers—*Protohertzina* of the portcullis jaws—are lowered within a cage on to the arena floor. As soon as the cage door opens, these hungry-looking creatures start to sniff around the circle, looking for lunch. It is only then that we notice that a third actor—our little trilobite, *Callavia*—has been tossed somewhat uncaringly across the circle and into the arena. The predators quickly catch sight of the trilobite, and start to advance. The audience urges little *Callavia* to flee. But there is no escape route, and the small creature is obliged to dig hurriedly downwards into the sand. Unhappily, she does not have time to dig quite deep enough to stop one of the predators from making a quick smash-and-grab. This it does, shaking her broken body victoriously from its spiny jaws. It is only then that we notice, in the spotlight, the rather desperate scratch marks left behind by the unlucky trilobite on the sandy floor (see Figure 11). These are the marks we geologists call *Rusophycus*.

Although we have cast this story as a kind of circus entertainment—or rather, by the end of it, a gladiatorial combat—something like each of these performances will have taken place, as urgent as a fight between lions and Christians in ancient Rome. Each was a matter of life and death. These circus acts also help to shine a light upon the suggestion, mentioned in the previous chapter, that one fundamental attribute of the Cambrian explosion was the evolution of the mouth. As we have already suggested, some organisms were content to swallow mud or seaweed. Such organisms will probably have included the makers of the meandering tracks of *Helminthoida*. Others, such as the makers of *Rusophycus*, tried to avoid being around at lunchtime, by burrowing into the sediment. That meant that predators would have to dig for them if they wanted to dine.

These scribbles left behind by animals on the ancient seabed are therefore some of our strongest clues concerning happenings near to the start of the Cambrian. That is why, in 1992, it was trace fossils and not small shelly fossils that were used to define the base of the Cambrian, at Fortune Head in Newfoundland.[106] Not only did protozoans such as *Platysolenites* seek the protection of skeletons. But soft-bodied worms started to burrow down into the sediment. Significantly, such worm wiggles can be seen worldwide in marine sediments of this age (see Plate 8 and Figure 12). In other words, the Cambrian explosion was truly global. And it seems likely to have been a real evolutionary event.

A major pattern is therefore beginning to emerge from the fossil record that we now need to ponder. This concerns the act of burrowing itself. It seems that Precambrian animals, that were previously content to creep about on the surface, found it strongly advisable to start digging deeper in the sediment,

Figure 12. The earliest evidence for the circus of worms in eastern England can best be seen in the eighteenth-century gravestones of Swithland Slate in Charnwood Forest, Leicestershire. Here the author (at right) points out trace fossil evidence for the Cambrian explosion to Canadian palaeontologist Guy Narbonne (at left).

and almost as soon as Day One of the Cambrian came along. Why was that?

One explanation for this digging, and for the new-fangled skeletons as well, is that the start of the Cambrian ushered in a kind of war of escalation between rival armies. Swords and arrows—jaws—on the one side were met with shields and helmets—skeletons—on the other. Battering rams were stalled by castle moats. Machine guns were trounced by tanks. And tanks were then halted by trenches—the burrows and trails we have just seen. Of all the hunches we have looked at so far, therefore, it seems that explanations involving intrinsic causes, like

Sollas's Gambit (and the Cambrian Cascade; see Chapter 4), are beginning to look at least as plausible as extrinsic causes (such as Daly's Ploy) for explaining the Cambrian explosion.

Calling Lyell's bluff

Is the apparent lack of animal fossils in the Precambrian—in Darwin's Lost World—a real phenomenon or a 'bluff'? Were the missing cards still in the deck (hidden in the rock record), in the hands of other players (perhaps hidden away in museums), or simply not there at all?

As we have seen, three main gambits have been put forward as answers to explain this pattern, which we have called Lyell's Hunch, Daly's Ploy, and Sollas's Gambit. Of these, Lyell's Hunch is the one followed by Charles Darwin in 1859. It predicted that the absence of animal fossils before the Cambrian would be found by us to have resulted from a poor fossil record. More recently some researchers, such as Richard Fortey, have predicted that the animal ancestors in the Precambrian will turn out to have been, for the most part, too small and delicate to leave a fossil record. But none of this fits too easily with the Circus of Performing Worms we have just seen. Animal tracks and trails are remarkably robust, and also tend to get sharper with time and weathering. Even the smallest of worms must surely have left behind some kind of spoor in the fossil record.

Daly's Ploy argues for a very different scenario. It predicts that the explosion of animal fossils near the start of the Cambrian should have been accompanied by evidence for dramatic changes in ocean-atmosphere chemistry. But that does not entirely fit the picture now emerging. For example, creatures that made the agglutinated tubes of *Platysolenites*, the tracks of *Rusophycus*, and the trails of *Gyrolithes* all seem to have been remarkably disinterested in carbonate chemistry. There is not a jot of evidence as yet that they had any need for waters that were becoming richer in phosphorus, in calcium, or in oxygen. That leaves us with Sollas's Gambit. This predicts that the fossil record should reveal to us the symptoms of an evolutionary transformation, perhaps an avalanche triggered by a number of innovations. For the present we can only guess at what those innovations might have been.

Whatever the ultimate cause of the Cambrian explosion, the result in evolutionary terms was so large that it not only raised the visibility of certain 'cards' in the game—the skeletal fossils—it also changed the rules of the game itself.[107] It was no longer possible to hang around unprotected on the seafloor. A change within the system had taken place that brought some new and alarming 'kids on the block'. And two of the new solutions for dealing with this menace of the mouth—skeletal armour, and underground burrows—may have led towards surprising changes on the planetary surface, as we shall see.

CHAPTER SIX

A MISTAKEN POINT

The Cape Race road

The road that leads from Portugal Cove to the lighthouse at Cape Race can feel like the most desolate place in the North Atlantic. It follows a line of telegraph poles that march in single file across forty kilometres of rolling bog and moss. And the wind across this plateau is so strong and constant that no tree can gain a foothold. On the day we arrived in 1987, a thick grey mist had rolled in from the sea, turning daylight into appropriately Celtic twilight—for this is the Irish corner of Newfoundland.

Our conglomerate of geologists had come in search of that famous slope of slate known as 'Mistaken Point'. The name recalls a sad moment, a century ago, when the captain of steamship *George Washington*—held at sea by poor visibility—mistakenly turned his wheel to starboard some 12 kilometres too soon, with tragic consequences for all passengers on board. But the hard grey rocks around Mistaken Point are more famous for another tragic loss of life. They preserve that mysterious moment in the history of the biosphere near the end of the long Dark Age of the Precambrian and before the Golden Dawn of the Cambrian.[108]

To see the fossils of Mistaken Point, we were obliged to leave the track behind, about halfway along the Cape Race road, and hike down a muddy path that winds towards the sea. Once the coastline was reached we could spy, all about us, huge slabs of dark green rock jutting out to sea at jaunty angles—some plunging down to the boiling waves, others tilted towards the sky. Not surprisingly, we sought out the latter kind, so as to avoid tumbling into the brine beneath. There were a half dozen or so bedding planes, each a good ten feet or so above the Atlantic breakers, and each reached by a short scramble down the cliff.

As we arrived at the fossil site, our chatter was interrupted by a great roar. Yet another wave had ended its journey across the Atlantic. It came thundering into a rock beneath my feet, throwing up columns of salty spray that fell back upon the slab with a sharp *slap*! How the fossils survive this constant onslaught of the waves—let alone the winter's ice and snow—is merely the first marvel. But even stranger things can happen here before dark, too. As the sun descends, the surface of the rock awakes.

A trick of light takes place at sundown. Looking down on the bedding from rocks above gives an impression of snorkelling above an ancient sea bed, some 563 million years old, completely covered in creatures. Hundreds of fossils unfold, preserved in high relief. Here, then, are some of the most visually stunning fossils on the planet. Surely, this is one of the greatest free shows on Earth.

These creatures have been likened, fancifully of course, to the flower beds of an English country garden, planted with 'hart's tongue' ferns (*Charnia*), clusters of ornamental 'lettuces' (*Bradgatia*) and scatters of giant 'daisies' (*Ivesheadia*) across the bedding. The fossils are not remotely related to modern flowers, of course—flowers did not evolve until the time of the dinosaurs, in the Cretaceous. Instead, these are the earliest known candidates for our animal ancestors—some of the missing fossils of Darwin's Lost World. But are they really the remains of animals? Time, then, for us to take a closer look.

The Spindle Animal

Perhaps the most striking and abundant of the fossils on this surface is the so-called 'spindle animal', recently named *Fractofusus*. Several hundred impressions of these spindles occur on a single bedding plane, some up to about thirty centimetres long (see Plate 9). But it is the astonishing detail of the preservation which delights the eye. The spindle shapes look as though they have been cut with a scorper tool from the hard rock by a master engraver. This kind of preservation, moulded like an intaglio in a gemstone, or a human footprint impressed in beach sand, shows it was the undersurface of this mysterious organism that we are seeing here. Planting our noses closer to the bedding plane, it is easy to see that each spindle fossil is made up from dozens of tiny fern-like bushes, ranging from larger clusters in the middle towards smaller clusters at each end. *Fractofusus* therefore looks a bit like a Victorian herbaceous border, provided with neat little rows of shuttlecock ferns. And just like ferns, their fronds resemble the famous fractals of the Mandelbrot set, hence their Latin name of *Fractofusus*, meaning 'the fractal spindle'.[109]

Scattered here and there among the spindles are impressions called *Bradgatia* (see Figure 13). This fossil resembles a lettuce that has made a bid for freedom from the vegetable garden and decided to go it alone. Some of these can look as neat as market produce while others have seemingly gone to seed, being both large and rather straggly. A closer look at the bush of *Bradgatia* shows that, like *Fractofusus*, it was constructed from units, each of which

Figure 13. The lettuce-like impression called *Bradgatia*. This drawing was made by the author using laser analyses of the type specimen from Charnwood Forest in England. It is about 565 million years old. It clearly shows the tightly clustered leaves, each built up from numerous tiny *Charnia*-like elements. This fossil is some 40 cm long.

resembles a single fern frond. Indeed, something like these fronds can also be found preserved singly, when they are known as *Charnia* (see Plate 10). Like a handsomely groomed ostrich plume, *Charnia* comes provided with alternating veins and complex subdivisions. In England, some specimens of this famous fossil can grow to nearly a metre length, but most are of fist size or smaller.

Ring of truth

Discovery of these strange fossils emerged in a lengthy and roundabout way. The first structure was spotted on a bedding plane in central England, in 1866 and reputedly as far back as 1844:

> Whatever the cause may have been, there are at present no known [fossiliferous] markings on any of the slate-rocks of Charnwood, except a few curious and regular furrows on the plane of bedding, in one of the quarries at Swithland.

> Considerable attention was paid to these by Mr. J. Plant of Leicester immediately after their discovery several years ago, when the face of the quarry was first bared. Casts were also taken by him of the most remarkable specimens. Mr Plant imagined them to be coral. Professor Ramsay suggested they might be sea-weeds. They are concentric ridges and furrows—several in number in each case. Three or four good specimens were seen, and others were less perfect on the same exposed face of rock.[110]

These large ring-like structures were up to a foot across. For more than a century, however, nothing much happened with their study. The great-and-good of the geology world, including Thomas Bonney of University College in London, had dismissed 'the ring' as an inorganic structure back in the 1890s and all interest therefore evaporated. By 1947, a major tome on the geology of Charnwood Forest made no mention of these 'fossils' at all.[111]

We then move forward to 1956. It was in that year, a schoolgirl called Tina Negus first stumbled upon strange markings while picking blackberries alongside a steep rock face at the edge of what is now Charnwood Golf Course in Leicestershire. When Tina reported this odd discovery to her schoolteachers, they reputedly refused to believe her.[112]

On 17 April 1957, two schoolboys, Richard Allen and Richard Blachford, were climbing the same rock face, when they also noticed a strange marking, something like a fern frond. They called in another schoolboy, Roger Mason, who in turn went to report it to Dr Trevor Ford in the Geology Department of nearby Leicester University. Trevor also doubted such a far-fetched story. So poor Roger went back to muster his father, who had seen the fossil, and together they managed to get Trevor out in the field to view this strange marking. It was soon clear that they had discovered one of the most beautifully preserved, and hence most convincing, of all known late Precambrian fossils, later called *Charnia masoni*.[113] Sir James Stubblefield—Director of the Geological Survey but affectionately known as 'Stubbie'—jumped aboard a fast steam train to come and verify the find. At his behest, huge 200 kilogram blocks were then levered out of the cliffs by local quarrymen familiar with slate, and the blocks were then arranged for public display in Leicester.

Trevor Ford, to his credit, was content to be cautious about the nature of this fern-like organism called *Charnia*. Nothing like it had ever been seen before. He therefore suggested that it was some kind of seaweed impression, perhaps because of its resemblance to a modern seaweed called *Caulerpa*.

The seaweed that never was

Trevor Ford's cautious views were soon to be swept aside, however, by a much more glamorous idea—that *Charnia* and its relatives were nothing less than the ancestors of the earliest animals. That idea was arresting because the quest for our earliest animal ancestors constitutes a 'holy grail' in science. From 1959 onwards, the late Martin Glaessner was to play a kind of King Arthur figure in this quest.

Martin Glaessner made a rather precarious start as a scientist. I learned something of this when he invited me to dinner as his guest at Fortnum & Masons in London back in 1983.[114] We chatted about those things we shared in common—we were both authors of textbooks on microfossils, explorers of foraminiferid evolution, and head-over-heels in love with the earliest animal fossils. Over a lobster salad, the other Martin leant over to tell me the story of how he had begun his professional life in Vienna, Austria. Young Glaessner then moved on to study microfossils down a microscope in Moscow. The war saw him suffering from a long spell of scarlet fever, during which he found himself isolated in a Russian military hospital with nothing much to do. Glaessner was not prepared to let the devil catch him idling. So he set about writing a text book on microfossils, by scribbling down notes on borrowed hospital paper. Page after page of notes had then to be boiled—to kill the contagious fever—before they could be sent to his wife in Vienna for typing.

During the later stages of the Second World War, he moved to Australia, where his interests were quickly drawn towards the strange and ancient fossils of the Ediacara biota. And it is with these that our story of animals in the Precambrian really begins in earnest.

A circular argument

The famous Ediacara fossils of South Australia were actually discovered not by Martin Glaessner but by an Australian mineral surveyor called Reginald Sprigg whilst preparing a report on defunct mines in the Flinders Ranges.[115] Whilst eating his lunch one day, Reg idly turned over a slab of sandstone and was immediately struck by interesting markings hiding on its undersurface. The sheep station on which he made the discovery was called by its old aboriginal name of Ediacara (pronounced Ee-dee- *a*cra). And this is still the name used to describe not only fossil assemblages of this kind around the world but it is also the name for the time interval—the Ediacaran period—during which they flourished.[116]

A strange circularity in thinking then began to emerge. Sprigg's Ediacaran fossils actually occurred some hundreds of metres below the earliest known trilobites in this area. But by 1947 it was widely believed that the Precambrian did not contain large and visible animal fossils. That meant that any rocks with such fossils must, by definition, be regarded as Cambrian!

But there was worse, too. Although he had discovered *Charnia*-like fossils, and a host of other odd-looking beasts, he could get no one to support his ideas. For example, when he showed these strange markings to Sir Thomas Playford, Premier of South Australia, the latter exclaimed: 'Give me some rich copper-lead ore, any day!', presumably because Australia depended then, as it does now, on minerals for the growth of wealth.[117] And when he talked about them at a major Australian science meeting, and even at the International Geological Congress in London in 1948, he was to hear them airily dismissed as 'fortuitous inorganic markings'. His report to *Nature* was, of course, promptly rejected and returned as holding limited scientific interest. No surprise, then, that poor Reg Sprigg gave up on his fossils and went looking for oil.

From 1957, Martin Glaessner therefore had a relatively free hand to set about solving the mystery of the enigmatic Ediacara biota to his own satisfaction. Indeed, he devoted much of his life to their description and interpretation.[118] By the time that Trevor Ford was describing *Charnia* as a fossil seaweed from England, Martin Glaessner had come to the conclusion that *Charnia*-like forms in Australia were not seaweeds at all. Instead, he suggested they were the remains of feather-shaped soft corals like those alive today, called 'seapens'. In other words, Glaessner regarded *Charnia* as a member of that seemingly simple group of animals known to zoologists as 'cnidarians' on account of their possession of little stinging cells called cnidocytes. Anyone who has blundered into a jellyfish or a fire coral knows about these cnidocytes—they are used to stun and capture their animal prey. As we have seen in Barbuda, they leave a prickly sensation on the skin, like that of falling into a stand of stinging nettles or poison ivy. These stings are strong enough to stun and kill a small animal.

Glaessner's 'decoding' of the Ediacara fossils seemed, for decades, to be a great revelation—providing evidence for the long-awaited ancestors of the Cambrian explosion. To use the terminology of the time—they were regarded as evolutionary 'missing links'. The earliest fossils, it was thought, could be compared with some of the most primitive animals still alive today. From 1958 to 1984, Martin Glaessner and Mary Wade neatly pigeonholed dozens of curious fossils from the Ediacara Hills into modern taxonomic categories.[119] Each new description revealed yet another, long awaited, progenitor of the modern animal world—the first jellyfish, the earliest worm, the ancestors of crabs, and the precursor of sea urchins. Accolades, prizes, and medals soon started to roll in.

And a mistaken conception

At that time, therefore, the continuity of fossil lineages from the Ediacaran period into the Cambrian seemed assured.[120] When the Mistaken Point fossils were discovered in 1969 on the other side of the planet by a Master of Science student called Shiva Balak Misra, it was therefore natural for geologists to think that this was merely another example of Glaessner's jellyfish world. But this view was not to last. The ideas of Martin Glaessner were about to be turned on their head.

The gauntlet was first thrown down by an eccentric German palaeontologist from Hamburg called Hans Pflüg, who had been working on similar fossils from Namibia in southern Africa.[121] Two fossils, in particular, stand out. One was called *Ernietta*, which looked like the crown of a Homberg hat pushed into the sediment. The other was banana-shaped *Pteridinium*. Hans Pflüg could not see in his Namibian fossils the representatives of modern jellyfish and worms that Glaessner saw. Instead, he argued that forms like *Ernietta* and *Pteridinium* were actually the remains of myriads of single-celled organisms (perhaps amoeba-like) arranged within a strangely shaped colony. They were therefore distant ancestors rather than fully fledged animals, and he called them the Petalonamae—the petal-shaped animals of Namibia.[122] It has to be said that Hans Pflüg's ideas about the Petalonamae were not well received by the scientific community. Instead, they were dismissed as little more than eccentric musings.

The views of Glaessner were about to come in for a proper pounding, though, from another geologist called Dolf Seilacher. Like Pflüg, he was also working in Germany, but in Tübingen. Seilacher argued that few, or maybe none of the Ediacaran creatures were ancestral to the Cambrian explosion. Instead, he proposed that they were uniquely quilted organisms, each constructed along the lines of a miniature air-bed. In his bold view, these *vendobionts* as he called them, were a kind of failed experiment, that died out at the end of the Precambrian. Pflüg and Seilacher had opened up an Ediacaran can of worms.[123]

Back to the Mistaken Point

In 2002, the geologist Guy Narbonne invited me to look yet again at the Mistaken Point biota, which had by this time been securely dated at between 575 and 560 million years old (see Figure 14). As we have seen, the Glaessner view, of some kind of sunlit Ediacaran coral garden, was starting to come under pressure. A new and fascinating set of problems was also beginning to emerge as well. In Newfoundland, the fossils were showing signs of having once lived on the seabed in very deep water, far from the reach of sunlight. The fronds could not, therefore, be the remains of algae—as once thought by Trevor Ford—nor of algae-bearing animals—as later thought by Mark McMenamin of Massachussetts, because algae need to live in sunlit waters.[124]

It was becoming hard to avoid the conclusion that *Charnia* and its relatives were not feeding upon sunlight and carbon dioxide in the manner of modern plants or algae. Could it be that *Charnia* was feeding on other organisms, a bit like soft corals perhaps? There are, after all, lots of soft corals that look a bit fern-like, especially seapens. Guy Narbonne and his student Matthew Clapham thought they could see a kind of ancient ecology in the rocks at Mistaken Point, with some of the longer fronds having fed high in the water column like modern sea fans and corals, and other lowly ones having fed at the level of the sediment, like modern clams.

But a further and equally fascinating problem then began to haunt us. Corals have a mouth, but nothing like a mouth has ever been demonstrated in any Ediacaran fossil. Without a mouth or a gut, it would have been difficult for these organisms to feed directly on organic matter.

The Darwin Centre

By the turn of the millennium, Martin Glaessner's idea was still highly popular: that *Charnia* and its relatives were some kind of ancient soft coral, like a seapen. To test this possibility I therefore went, with Jon Antcliffe, in search of some modern seapens.

As luck would have it, we found that a fine collection had just been made available to the public in London—at the Darwin Centre of the Natural History Museum.

From the outside, the Natural History Museum in London can seem like the epitome of a Victorian Gothic building—all arcs, aardvarks, spires, spirals, and zebras informally jumbled together in a great Noah's Ark of stone, steel, brick, and glass. From the inside, this impression is yet further heightened by the lofty atrium, leading the eye towards a staircase that ironically recalls the ascent of life itself. I say 'ironically' because Sir Richard Owen—Darwin's mortal enemy—now stands forbiddingly at its top.

For a century or so, this great hall in South Kensington has been home to some of our favourite dinosaurs. One of these—called *Apatosaurus*—is an epitaph to the Victorian philanthropist Andrew Carnegie. So pleased was Carnegie with this old fossil that he ordered about a dozen huge replicas to be put on display around the world. Mercifully, the modern management has not yet tinkered with this entrance hall, allowing it still to entrance us.

The object of our quest lay, however, at the back of the Darwin Centre and hidden from the public gaze. Each floor of the Centre is devoted to different kinds of animal, vertebrate and invertebrate, preserved and pickled within spirits of one kind or another. Seapens dwell on the fourth floor. To reach them, we had to pass a vast gallery filled with row upon row of neatly catalogued beasts: sponges, coral, jellies, squids, cuttlefish, worms of this tribe and that; long jars, short jars, little shiny squat jars—all lovingly crafted by Victorian glaziers to hold these tiny, or not so tiny, treasures.[125] Then Jon's eye was caught by a jarful of scarlet seapens, looking like the ruffled feathers of a cockatoo, called *Pennatula phosphorea*. We could see that *Pennatula*, like its relatives, grew in stages by the addition of soft little polyps near to the base of its stem. The polyps near the top of the frond looked stiff and old. *Charnia* and its relatives, however, seemingly grew by the addition of little segments at the very tip of the frond. It was the segments at the base that loomed largest, suggesting that they were the oldest.[126]

We could also see that these modern seapens are provided with a highly muscular foot that is known to be used for burrowing. Indeed, seapens spend most of their time hiding in their burrows, as Charles Darwin himself was whimsically able to observe in 1835: 'At low water hundreds of these zoophytes might be seen, projecting like stubble . . . a few inches above the surface of the muddy sand. When touched or pulled they suddenly drew themselves in with force, so as nearly or quite to disappear.'[127] But no burrows of any sort have ever been confirmed within the Mistaken Point biota. Whatever it was, *Charnia* was not like one of Darwin's seapens.[128]

The paradox returns

These problems were starting to suggest that both *Charnia* and *Fractofusus* did not feed on sunlight. Nor did they feed like seapens by catching prey. Instead, they could well have fed by absorbing chemicals or food matter directly from the water column, much like the vent worms living around 'black smokers' in the deep sea today. Unhappily, however, such a simple interpretation was to be shaken by yet another direct observation on the fossils themselves: *Fractofusus* and relatives are closely impressed into the mud, in ways that suggest that they actually lived 'face-downward' in the mud.[129] When all these familiar feeding strategies—feeding with a mouth, feeding from algal symbionts and feeding by absorption from the water column—are discounted, rather few options are left to us. As Sherlock Holmes once said to Dr Watson: 'How often have I said to you that when you have eliminated the impossible, whatever remains, *however improbable*, must be the truth?'

An increasingly discussed possibility for the strange Ediacara biota is therefore this one: that many of these creatures were actually absorbing food materials directly from the mud itself. This remarkable conclusion has moved some recent researchers, such as Greg Retallack and Kevin Peterson, to reject the traditional Glaessnerian interpretation of true animals being preserved within the Ediacara biota. That is because animals, by definition, typically feed largely by ingestion—by means of swallowing food via a mouth—rather than by absorption. Now there is one group of organisms that needs neither mouth nor anus, and feeds instead by absorption: moulds and mushrooms, both of which belong to the Fungi. Interestingly, molecular research suggests that Fungi may indeed be rather close to the ancestral line of modern animals. But few scientists are inclined to accept this view just yet. That is because the complex, fern-like pattern of construction looks nothing like that seen in any known group of living fungi.

The wrong pizza

One of the oldest fossil-like structures seen on the bedding planes at Mistaken Point looks like a flat-pan pizza, complete with dimples of pepperoni and toppings of onion rings. There are many dozens of them scattered across the bedding planes alongside spindles, bushes, and fronds. It seems clear that they had a growth programme because they can range in size from several centimetres to nearly a metre across. Like a cheap pizza, however, their biological origins can seem a little hard to detect.

Such fossils were first described by Helen Boynton and Trevor Ford from rocks in Charnwood that may be more than 600 million years old. Helen has a marvellous knack for seeking out such strange markings. For her, these old rock surfaces are immensely tactile, and she reads them like a blind person reading a Shakespeare sonnet written in Braille. Some very strange ideas have been put forward for the biology of these pizza discs, more correctly called *Ivesheadia lobata* (see Figures 14 and 15). Some have suggested they were balloon-like organisms of uncertain kind, tethered to the seafloor by strings. This illusion refers to the fact

Figure 14. Palaeontologist at work. The author is shown mapping out an Ediacaran fossil impression in the field near Mistaken Point in Newfoundland, in rocks some 575 million years old. Such mapping and measurement precedes the making of high resolution casts. These casts are then used by us for computer-assisted imaging techniques, including laser scanning. Such work allows the detailed reconstruction of complex fossils like those of *Bradgatia*. Image taken by Duncan McIlroy.

Figure 15. The pizza-disc-like impression called *Ivesheadia*. This drawing by the author was made by means of image analysis of fossils and casts from Mistaken Point in Newfoundland. It occurs in rocks more than 570 million years old. It shows the rather disorganized arrangement of numerous *Fractofusus*-like elements nestling within the dimples of the pizza disc. This fossil is some 40 cm across.

that some of the pizzas seem to have long threads coming out of them, up to several metres long. Others have regarded them as possible filter-feeding colonies, perhaps even poorly preserved colonies of sponges. Unfortunately, no sponge spicules, pores, or other parts of the 'sponge identikit' have yet been found. Instead, as Duncan McIlroy has observed, these pizza discs often seem to contain the decomposed remains of something a little more familiar. Each dimple in the pizza disc contains little clusters of *Fractofusus*-like frondlets. In other words, the pizzas could have been large and complex colonies that were living on the seafloor in much the same way as the spindles. Unlike most of the spindles, however, these strange *Ivesheadia* colonies were broken down into something resembling Mozzarella cheese by the processes of decay on the Ediacaran seafloor.

What-the-Dickins?

The oddities we have looked at so far come from older rocks around the North Atlantic. Surely there must be signs of the animal ancestors in Martin Glaessner's own happy hunting ground of South Australia? These younger rocks were laid down a mere 10 million years or so before the start of the Cambrian explosion. Do they contain a solution to the problem of Darwin's missing animal fossils?

One of the commonest fossils to be found in the Rawnsley Quartzite of South Australia is called *Dickinsonia*. This somewhat resembles the impression of a big fat human fingerprint—an ovate disc that is neatly divided into dozens of segments by ridges that radiate away from a mid line (see the title page, Plate 11 and Figure 16). As with most of the other fossils from Australia and the White Sea of Russia, *Dickinsonia* is usually found as a negative mould on the under surface of a bed. Conditions for hunting the fossils are therefore very different from those seen at Mistaken Point, where we need only to walk over a bedding plane to see the fossils

Figure 16. What the dickins? A suggested reconstruction in life of the impression fossil *Dickinsonia*, drawn by Leila Battison and conceived by Jon Antcliffe. This fossil, which first appeared in the fossil record some 555 million years ago, ranged from a few centimetres to nearly a metre across. The makers of these strange impressions left behind no clear evidence for a mouth, anus or gut, nor did they have a proper bilateral symmetry.

laid out before us. In Australia, it is necessary to turn large rocks slabs over or, worse still, slither under a rock overhang in the hope of seeing fossils (see Plate 12). Much harder work.

The smallest impressions of *Dickinsonia* are about the size of a garden pea, while the largest can be of stupendous size—up to a metre across. Martin Glaessner was minded to see in these curious fossils the ancestors of modern worms. In particular, he was seemingly struck by their resemblance to a modern polychaete worm called *Spinther*. Unfortunately, however, we now know that this little worm, only about 5 mm long, is a highly specialized parasite that feeds only upon certain sponges—meaning that its simplified form is unlikely to be a sign of great antiquity. There are many other problems with this worm hypothesis, too. There is no convincing evidence for a head, a mouth, a gut, or even a humble anus in *Dickinsonia*. Worse, as Russian fossil expert Mikhail Fedonkin has shown, the segments of *Dickinsonia* typically alternate along a mid line, producing what is known as a glide plane of symmetry.[130] No animal either alive or dead is known to have had such symmetry. Except, of course, for *Charnia, Bradgatia, Fractofusus* and *Ivesheadia*.

Bruce Runnegar from Los Angeles has shown that *Dickinsonia* could grow to be as large and thin as a table cloth, so it is no surprise that it rumpled up and folded over very easily.[131] It is hard to imagine, though, how such a thin and flexible organism could ever have moved across the seafloor. One solution, perhaps, is that it didn't move at all but simply sat on the seabed, absorbing goodness from the ooze beneath. There are signs, though, that *Dickinsonia* could bounce across the seafloor, leaving behind a set of ghostly impressions in the mud. But sometimes, fact is stranger than friction. The simplest explanation for this set of impressions is a phenomenon we have all observed with amusement on a windy day—a sheet of newspaper happily flopping along the pavement. Something like this seems to have happened on the Ediacaran seafloor, perhaps under the influence of waves or currents.

An upside-down argument

At first sight, the narrow body of the fossil called *Spriggina* looks a bit more convincing as a worm ancestor than does the table cloth of *Dickinsonia*. The body is about the size and shape of a little finger and it is likewise divided by transverse ridges and grooves. The place of the 'finger nail' is then taken by a crescentic 'head'.

Martin Glaessner was clearly struck by the similarity between this strange fossil and a living tropical worm called *Tommopteris*. This 'Pololo worm' can swarm in huge numbers at the surface of the Pacific ocean, especially at certain points in the lunar cycle, which it uses as a time check for the dating game. *Tommopteris* is therefore an appealing search image for *Spriggina*, and this comparison has become the accepted nostrum for decades. Accepted, that is, until, Dolf Seilacher decided to turn the argument about *Spriggina*, and the fossil itself, completely upside down.

So besotted were we, said Dolf Seilacher, with this Pololo worm model that we had neglected to look at *Spriggina* orientated the other way up. In other words, we should look at it with its crescent at the bottom of the beast—like an anchor—rather than at the top of the beast—like a head. When turned arse-over-apex in this way, *Spriggina* becomes magically transmuted into a *Charnia*-like beast. Moreover, it is also provided with that curious diagnostic criterion of the vendobionts, a glide plane of symmetry. In other words, scientists had arguably been falling for a kind of Inkblot Test.[132] What we see in our fossils therefore depends utterly on our points of reference—in this case, our hidden assumptions about polarity.

We are required to be able to tell our arses from our apices. Clearly with *Spriggina*, we cannot yet do so.

The fib in Fibonacci

It was once widely agreed that we could see snapshots of jellyfish, worms, and arthopods in the fossils from the Ediacara sheep station. Seen from that perspective, the picture only awaited evidence from other major groups of animals without backbones such as, for example, the echinoderms, a big group that includes starfish and sea urchins. It is at this point in the story that a fossil called *Tribrachidium* is usually pressured into service.

Tribrachidium is a circular impression about the size of a large coin, within which can be seen three 'arms' curving away from the midpoint, rather like the three-spoked Celtic wheel-of-legend, called a triskele. Comparable fossils called *Arkarua*, provided with five arms rather than three, were later discovered by Australian geologist Jim Gehling.[133] It has for long been a commonplace to regard these fossils as relatives of living starfish and sea urchins—echinoderms provided with a fivefold pattern of symmetry. But this beautiful hypothesis conveniently overlooks two slightly ugly little facts. First, neither five- nor threefold symmetry can be found in the earliest undoubted echinoderm fossils from the early Cambrian. This suggests that such symmetry patterns were a later feature. Second, both five- and threefold symmetry are common in the living world today—in creatures that range from protozoans such as *Quinqueloculina* (which has fivefold symmetry, see Chapter 1) and *Triloculina* (threefold) to the five-petalled wild rose and the three-petalled lily. Both arise because the numbers 3 and 5 are part of the famous Fibonacci series: one, two, three, five, eight, thirteen and so on. This series thrives within nature because of a very simple fact. Five divided by three—or thirteen divided by eight—generates that deeply irrational number called ϕ, which itself allows the stable packing of self-similar elements around an axis. Wherever folding and self-similarity are in evidence in nature, then patterns of three and five are likely to emerge.[134] Curiously, though, this means that the possession of three or five 'arms' tells us rather little about the *biological relationships* of an organism, be it a dog rose or an *Arkarua* disc. Happily for nature, but unhappily for echinoderm-fanciers, these numbers cannot be a definitive characteristic for an animal phylum, only a secondary and emergent one.

The most parsimonious explanation for *Tribrachidium*, therefore, is to see it as a basal impression left behind by some kind of *Charnia*-like creature. In support of that view, we can now point to types of 'Charniomorph' that were provided with threefold vanes. These can be seen in the remains of fossils called *Charniodiscus, Rangea, Pteridinium*, and possibly in five-flanged ones called *Swartpuntia*.

Farewell to jellyfish

By far the most abundant impressions on bedding planes in the Ediacara Hills are circular fossils, now commonly placed within the genus *Aspidella*. These circles range in size from little blobs, like nipples, to discs the size of dinner plates. The latter are often provided with nested rings that somewhat resemble the wrinkles on the skin of a rice pudding or, of course, the wrinkles on the 'bell' of a jellyfish.

The idea that these discs were the remains of jellyfish, like those seen stranded on a beach today has caught the popular imagination. That is why they are still called 'medusoids'. And that is why we can still enter our national museums and see dioramas with schools of diaphanous jellyfish pulsing through the ocean above an Ediacaran seafloor of rippled sand. Indeed, my own awareness of fossils was awakened by a poster of such fossil jellyfish pinned up in my classroom in 1954.

It was to prove a trifle inconvenient, therefore, that Trevor Ford in England had reported jellyfish-like structures attached to the base of *Charnia*-like fronds as far back as 1958. For decades, people therefore blanked out of their minds a most inconvenient truth: that the jellyfish-like discs of the Ediacara biota are not 'jellyfish-like' at all. They typically lack tentacles, gonads, and other bits of jellyfish tackle. Instead, many medusoids have proved to be part of the apparatus of Ediacaran fronds, and are now called 'holdfasts'. But even the term 'holdfast' for discs and medusoids is a bit misleading because it is just a guess. There is little evidence as yet that they were truly capable of 'holding the creature fast'—they seem too thin and flat. Other so-called 'jellyfish' impressions of Precambrian age have turned out to have an even less inspiring origin, such as microbial colonies, gas bubbles, mineral growths, or scratch marks left behind by seaweeds circling about on their holdfasts in the tidal ebb and flow.[135]

One of the last bastions for a world with Precambrian jellyfish was arguably the evidence provided by *Kimberella*, a fossil first described from the Ediacara hills of South Australia and recently reported from the White Sea of Russia (see Figure 17). Looking as it does, somewhat like a tiny plastic parasol squashed sideways in the mud by a strong sea breeze, most scientists were for long agreed that this was the impression of some kind of 'box jellyfish'. But even this evidence from *Kimberella* has now fallen prey to revision. Mikhail Fedonkin and Ben Waggoner have together speculated that the 'parasol' resembles not a jellyfish bell but the

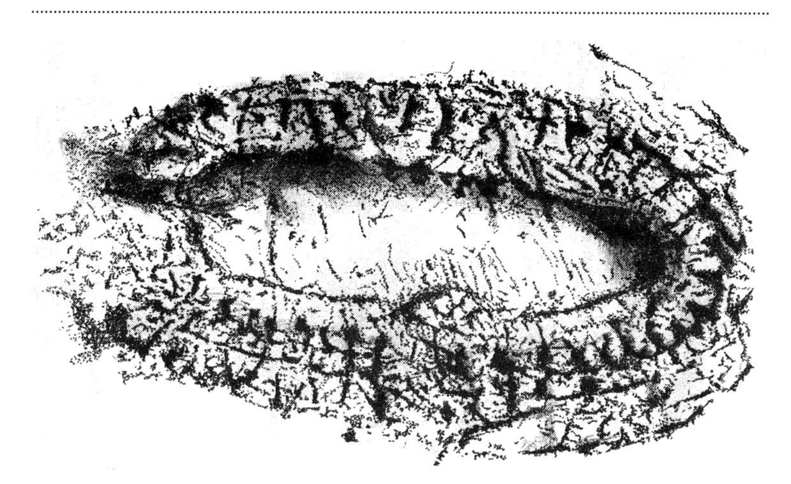

Figure 17. The jellyfish that wasn't. The impression of *Kimberella* is here shown in a scientific drawing by the author from 555 million-year-old material of the White Sea in Russia. These little fossils are usually about 1–3 cm in length. The frilly pattern around the edge has been compared with the mantle edges of modern molluscs, while the tubular structure at the left end has been taken to suggest some kind of feeding proboscis.

mould of a sea slug's body, provided with a skirt-like mantle and a slug-like foot.[136] Some would even go further. Little ridges on the same bedding plane have been used to argue that *Kimberella* scraped away at the seafloor in search of food using a specialized set of molluscan teeth, collectively called the radula.[137] But a much less exciting, and therefore less examined, explanation is that these ridges are simply partial moulds left behind by microbial filaments, by protozoans or even by decaying *Dickinsonia*.

From pit to blob

All this brings us back full circle to the first structures described from the Ediacaran interval—from the Longmynd in England. As we have seen J. W. Salter set off in 1855—with geological hammer in hand—to begin his three-week crusade in the search for clues to the origins of life.[138] On a bright cold winter's day, some 152 years later, a small group of us returned to this site, to search around the very same spot.

The Longmynd of Shropshire is a whale-backed hill all covered in heather and bracken, except where clear mountain streams have cut down through the slate to form deep V-shaped valleys. In the winter, the sun barely reached the bottom of these valley floors, so we found the grassy banks of the brook decked with frost. After a walk up the stream bed, we at last spied one of Salter's old eyries, high up on the northern slope, still bathed in morning sunshine. Here, the slatey green rocks seem to be covered in 'goose pimples'. A closer look shows that many are not round at all but two-lobed, like a cloven hoof print. Some bedding planes looked as though a herd of tiny deer, just a few centimetres tall, had stampeded across the ancient sea-floor. These are the strange fossils that he took, quite understandably, to be worm casts that he then called *Arenicolites didymus*.

Darwin was intrigued, and made a brief mention of this in the *Origin of Species*: 'Traces of life have been detected in the Longmynd beds beneath Barrande's so-called primordial [Cambrian] zone.'[139] But over the following century, opinions on Salter's fossils were to swing to and fro. Part of the problem was his claim that the splash-marks of ancient raindrops occur alongside *Arenicolites didymus* in the same beds. By 1967, when I was first shown these structures, the money was on the raindrops alone—nothing was biological. More recently, opinion has swung in favour of the fossils.[140] Indeed, a closer look by my student Richard Callow has revealed a remarkably preserved seafloor, complete with microbial filaments and seaweeds.[141] This texture—like a

Plate 1. Darwin's Great Dilemma. This is the study where he wrote *On the Origin of Species.*

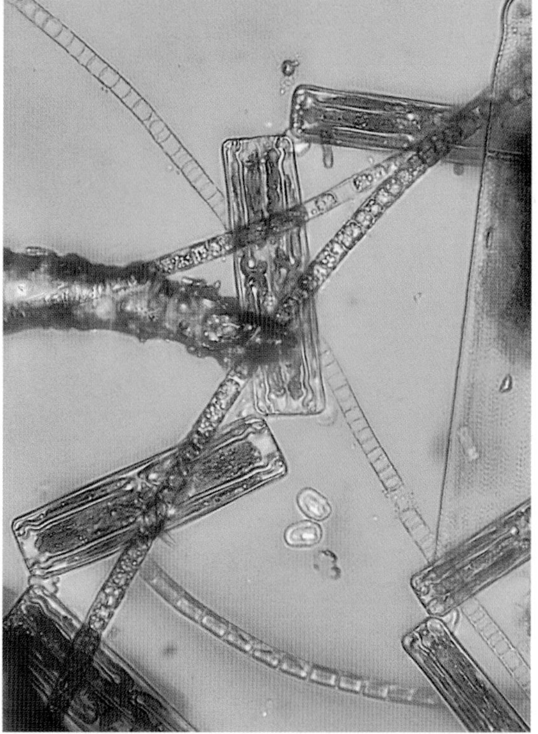

Plate 2. Up Cuffy Creek. Microbial mat like those in Barbuda, showing green algal filaments, box-shaped diatoms and a nobbly foraminiferid pseudo-pod. Field of view about half a millimetre across.

Plate 3. Bones of contention. The anatomy collection of Georges Cuvier in the Musée d'Histoire Naturelle in Paris.

Plate 4. The cliffs of Ulakhan-Sulugur along the Aldan River. The Tommotian fossils first appear in the red rocks. Russian geologist Alexei Rozanov along the shore, at right.

Plate 5. Fossilized jelly babies. The mountain slopes around Maidiping contain some of the earliest Cambrian phosphatized tissues.

Plate 6. Mongolian village with gher tents in the foothills of the Gobi Altai mountains. The surrounding hills contain very early mollusc remains.

Plate 7. The banner of International Geological Correlation Program Project 303, held up by members of the Mongolian expedition to the Gobi-Altai. Mongolian geologist Dorj Dorjnamjaa at left.

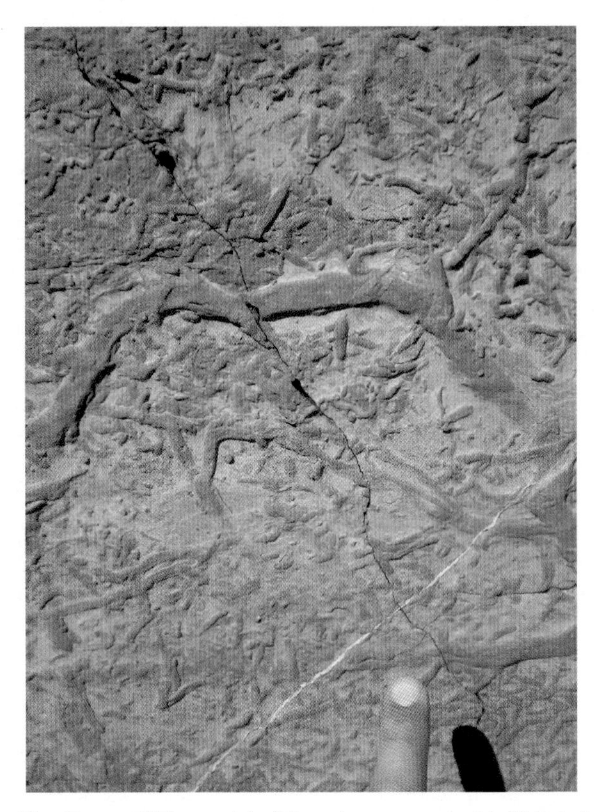

Plate 8. The Circus of Worms. A bedding plane covered with *Helminthoida* and other traces of animal activity of the kind that mark the start of the Cambrian explosion. From Fortune, Newfoundland.

Plate 9. The spindle-shaped fossil impression of *Fractofusus* on a bedding plane from Mistaken Point in Newfoundland, Canada. The fossil is about 20 cm across.

Plate 10. The leaf-shaped fossil impression of *Charnia* on a bedding plane from Charnwood Forest in England. The whole fossil is about 20 cm long.

Plate 11. The fingerprint-shaped fossil impression of *Dickinsonia* on a bedding plane of quartzite from the Ediacara sheep station, Flinders Ranges in South Australia. The fossil is about 5 cm long.

Plate 12. Searching for *Dickinsonia* and other Ediacaran fossils on the undersides of rocks in the Flinders Ranges of South Australia. Australian palaeontologist Jim Gehling (at far left) for scale.

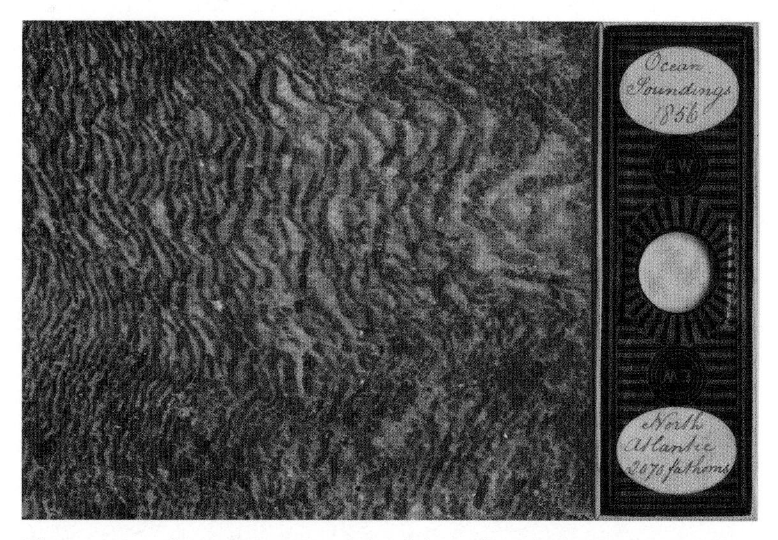

Plate 13. Eozoology. A Victorian slide of the earliest dredgings obtained from the deep Atlantic in 1856 is here placed alongside a Precambrian example of *Eozoon* from Quebec in Canada. The field of view is some 4 cm wide.

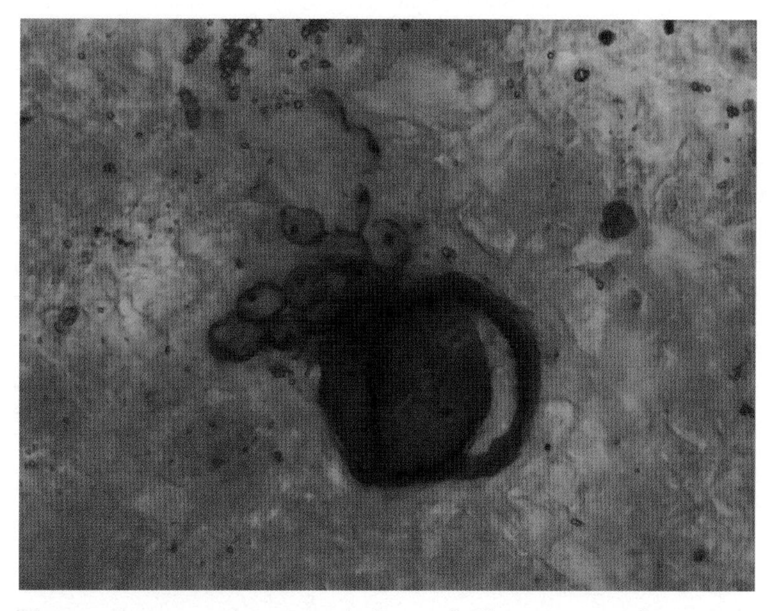

Plate 14. Bursting with activity. A cluster of cells germinating from a cyst, preserved in 1,000 million-year-old phosphate from the Torridon Lakes of NW Scotland. The cluster is about 0.2 mm across.

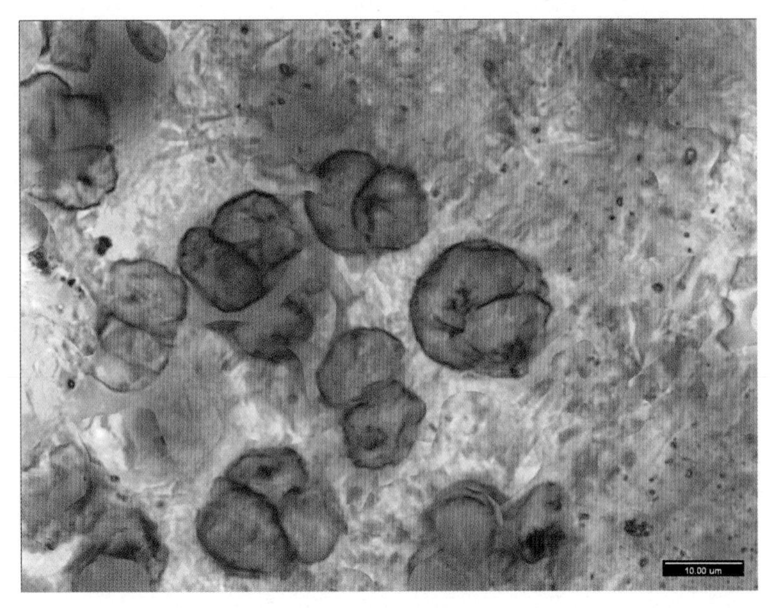

Plate 15. Pairs and small clusters of cells from the phosphatized Torridon lake beds. Each cell cluster is about 0.15 mm across.

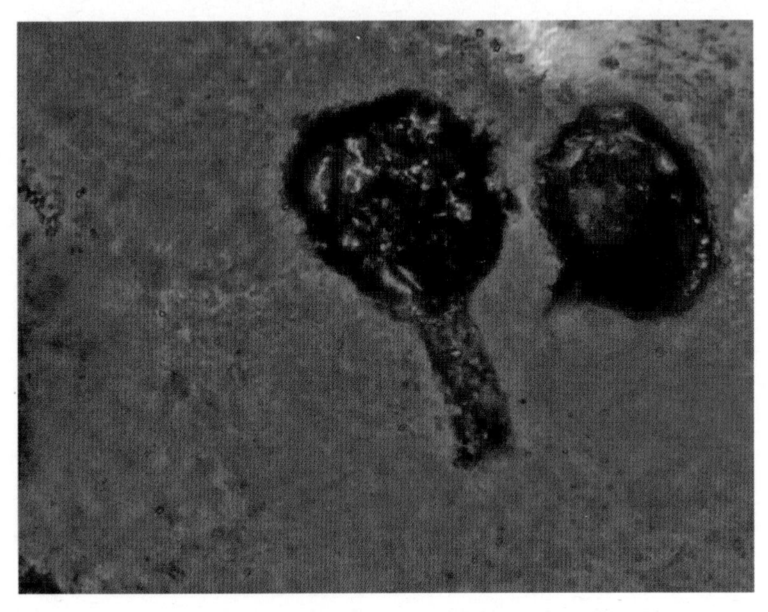

Plate 16. Two flask-shaped cells of protozoan form, from the phosphatized Torridon lake beds. The flask is about 0.6 mm long.

tatty old Persian rug coated with lumpy porridge—has turned up at about this level all around the world. Without knowing it, Salter had unveiled a vestige of our long Lost World.

The Mofaotyof Principle

Much of this puzzlement about the Ediacara biota arises because palaeontologists working on the early history of life adhere to what I like to call 'the Mofaotyof Principle'. Every science has its principles of operation—one has only to think of Lyell's Principle of Uniformity—and palaeontology is no exception here. The Mofaotyof Principle has a long and illustrious history, flourishing since the dawn of the discipline in the early nineteenth century. One of its first great exponents was even Darwin's great adversary, the palaeontologist Richard Owen.

Mofaotyof was not, as might be thought, a bespectacled Russian scientist from the days of the last Tsars. His name is fictitious. It stands for '*My Oldest Fossils Are Older Than Your Oldest Fossils*'. What concerns us here is the tendency among all scientists, and certainly among all journalists, to make their scientific claims as strong as they possibly can from the limited amount of material available. Indeed, the philosopher Karl Popper has ruled it a wise plan for scientists to make all their scientific claims as strong as they can—*so that the claims can be tested by others and, if needs be, rejected.* But strong claims are also the inevitable response to ever increasing demands for science funding. Not only is it necessary to win financial support for salaries, but there is also the pressure to publish and gain publicity for a career trajectory.

It is therefore important for us to ask ourselves, when assessing the claims for the 'oldest this' or the 'earliest that' to remember

Mofaotyof's First Golden Rule: *interpret the fossils to be as old as you can make them.* An example of this, which directly relates to our problem of Darwin's Lost World, concerns the discovery of strange markings in rocks from the Copper Belt of Zambia. These markings were reported in the eminent scientific journal *Nature* as evidence for the earliest signs of animal activity in the rock record. The author kindly brought them to show me soon after he had published this extraordinary claim. They had come from mine shafts deep beneath the land surface, and consisted of finger-shaped animal burrows infilled with red mudstone. The problem here was that rocks from which they were described were about 1000 million years old—almost twice as old as the other oldest burrows known from the fossil record. It later turned out that, yes, the structures were indeed animal burrows. But, no, they were not very old ones. These oldest claimants for animal life were actually the burrows formed by those modern 'white ants' we call termites. Those little insects live in colonies in hot regions that can be so dry that they need to dig galleries many tens of metres below the ground surface in search of water and nutrients. And that is what the Zambian fossil turned out to be—they are fossils but they are not really very old at all.

Deciding whether a fossil is old or not is becoming somewhat easier nowadays—there are plenty of tests that can be done. These include dating tests, like those at Oxford that demonstrated the Turin Shroud was yet another medieval fake. But Mofaotyof's Second Rule is even more cheeky. It says: *make the strongest, most biological and most newsworthy claim for your structure.* In other words, ask yourself: *what is the most exciting thing it reminds you of*? and avoid asking: *what is this structure really*? Everyone assumes, of course, that the structures being described are actually real fossils. But could it be that they are some other kind of structure that has come to resemble a fossil in the mind's eye? This Second Rule is like that of seeing the image of Mother Teresa in a currant bun, or the giant face of a man on Mars. Human culture is absolutely riddled with such cases of mistaken identity. Our arts, crafts, and media are essentially built upon such illusions. We come face to face with this paradox on our television screens every day. We may think we are seeing familiar faces. But they are not really faces at all—they are simply patterns on a screen. That may be taken to indicate a strong selective advantage in our being able to distinguish, very quickly and from limited information, the faces of friend and foe. But it often leads to interesting mistakes, such as in the Ink Blot (or Rorshach) Test used by psychiatrists. For example, when a psychiatrist asks you: 'What does this blot remind you of?' you may be tempted to answer 'the devil',' or a 'greedy pig'. But the only safe answer is, of course, 'an ink blot'.

In the century following the publication of Darwin's *Origin of Species*, several palaeontologists began to sail into the darkness we now call Darwin's Lost World. And all who ventured there seemed destined to run on to the rocks. Most of them suffered, too, to be swallowed up by ridicule or, even worse, by obscurity. Of these early adventurers, few had spent more time and energy in the field working on our question than did that great Antarctic explorer Sir Edgeworth David. What follows is a truly salutary tale, not about David and Goliath, but about David and the Gully Quartzite. I first learned of this sad story when Roland Goldring sent me a book he had found, that was published posthumously in 1936 by the Royal Society of New South Wales. Sir Edgeworth David, we know, had led a team to the true magnetic south pole, led the first expedition up Antarctic Mount Erebus, and he had even confirmed Darwin's theory of coral reefs on Funafuti. His name was, and therefore remains, sufficiently illustrious to allow us to view his private passion for Precambrian animal life with some kindly indulgence. He had, it seems, spent several field seasons between 1896 and 1926 searching for the earliest signs of animal life in South Australia. Just as we have done, he located those levels with the first trilobites and then worked his way backwards through time, and downwards through the rock record. Mile upon mile upon mile of rock he studied. And sadly, of course, he eventually found what he was looking for—giant arthropods and molluscs preserved in quartzitic sandstones that lay far below the level of the Cambrian. But nobody would believe him and his fossils. That is because they seemed to them rather like—and in fact actually were—ancient mud flakes that had simply tumbled about on the seafloor to make suggestive shapes. As far as I can tell, his steps towards the Mofaotyof Hall of Fame provide a lesson for us all on how such matters often tend to proceed:

1. *Make the claim sound truly authoritative:* something like this perhaps: 'No Australian scientist, whether palaeontologist, zoologist, or geologist who has taken the trouble to study the specimens, has any doubt whatsoever as to their being genuinely organic.' Or again, this might do the trick: 'Thus the time will surely come when forms, like those figured or now in our possession and to be figured later, will become almost as familiar to geologists and palaeontologists as are now many Cambrian fossils.'
2. *Be hungry for academic endorsement or acclaim:* Full marks go here to Sir T. W. Edgeworth David who had managed to collect the following: KBE, CMG, DSO, MA, DSc, ScD, FRS.
3. *Make competitors seem incompetent:* Sir Edgeworth David did his very best to cast doubt on the competing claims made by the

American Charles Doolittle Walcott, that *Beltina* from Canada was a pre-Cambrian arthropod. He was right, of course, it was nothing of the kind, not even an animal. But that is another story.

4. *Detail exhaustively the reasons why objects must be real fossils:* David and Tillyard then went on to list no less than six criteria across some seven pages of text, all of them questionable.
5. *Claim to have peerless experience and special gifts:* Like this for example: 'In regard to the study of these fossils workers in the old world will be at a disadvantage as compared with Australian geologists as they will lack that mass-action effect on the judgement which results from the examination of some thousands of specimens in the field, a valuable addition to knowledge based solely on the study of just a few specimens in a museum.' This is, of course, a classic 'put-down'.
6. *Become associated with other experts:* Sir Edgeworth David accomplished this by drawing in poor old R. J. Tillyard, an entomologist who was obliged to flounder in public, and was to find himself completely out of his depth.
7. *Keep out of the firing line:* Sir Edgeworth David did the best thing possible here. Like Sir Arthur Smith Woodward, one of the Piltdown dupes, he managed to die before his book was published. We may call this 'the master-stroke'.

Less baleful, perhaps, is another fine example of the Mofaotyof Principle, this time found in a book about Noah's Flood, written in about 1868 by Major General George Twemlow. This treatise goes by the snappy title: 'Facts and Fossils Adduced to Prove the Deluge of Noah and Modify the Transmutation System of Darwin with Some Notices Regarding Indus Flint Cores.' The Major General had gathered together sepia-coloured photographs of hundreds of different flint nodules from the Chalk of England, and invited the reader to see in them all manner of wondrous things. For example, he illustrated 'a pair of shoes, turned to flint, they are right and left foot, are bound together like a pair of new shoes (apparently of a child or of a Chinese lady)', found in a chalk pit near Guildford Railway Station. And 'a bird apparently on its nest, probably frozen'. In this case, the Major General shows us that he has scoured piles of flint nodules, numbering into the thousands, and picked out, from the mass of shapeless flints, all those which appear to mimic things in our own world.[142] We have all played this game as children—seen hideous faces in wallpaper patterns, or dinosaurs in cloud formations. The solution to this question therefore lies right before our eyes. We must never subjectively pick a suite of structures that seem to us to resemble living things. In other words, we must take an approach that looks at the *whole population* of structures.[143]

The final rule of the Mofaotyof Principle, it seems, is: *keep control of your story by limiting access to the fossils in question.* A beautiful story can be spoiled by an ugly little fact. If so, it might seem prudent never to let that ugly little fact reach the public gaze. Could such a gambit ever have happened in Academia? Do palaeontologists ever play the game according to the final rule of Mofaotyof? Well, yes of course, it has been known to happen. Every country has a faculty sitting proudly on their pile of fossils. Keeping key fossils hidden from view for many decades is a well-known but little-discussed problem. For the most part, this matters not a jot. To the winner go the spoils. But, where the earliest signs of life are concerned, scientists must also be careful to avoid any accusations of a conspiracy against thought.[144]

Many major palaeontologists have unwittingly tried their hand at the Mofaotyof game, myself included. The major scientific journals directly encourage it. The shortage of funding for science inevitably demands it. But without critical analysis, there is the danger of merely exchanging one set of creation myths for another. Happily, as this chapter shows, scientific ideas that are 'less fit' tend to collapse under the weight of adverse evidence, ultimately giving way to ideas that appear to explain more. Science operates according to Darwinian rules.

Dalradian worm poo

It is time to take a peek at the ways in which the Mofaotyof Principle may give us misconceptions about the nature of the early history of life. Some of these examples are mere curiosities. But others that have seemingly distorted our view for decades have yet to be expelled from the text books. I shall therefore begin with a confession of my own: the strange tale of 'the oldest worm poo'.

The following story arose out of an expedition I undertook to search for early life around two Hebridean islands off the west coast of Scotland, the peaty island of Islay—famous for its smoky whisky—and the rugged and remote island of Jura to its north. I first fell in love with these two islands in 1973, while working as a survey geologist aboard the drilling ship *MV Whitethorn* sailing out of Glasgow. We had been boring in the Malin Sea for days, recovering nothing but boring red muds from the equally boring (or so I thought) Triassic period. One night, though, the ship anchored off the island of Jura in a sea so calm that it seemed like a lake. The sun was setting behind 'The Paps' of Jura, two outrageously breast-shaped mountains of Precambrian quartzite. By the end of the voyage, my mind was set: I was going to give up my career as a survey geologist and pursue their story.

It was another six years before I finally set foot on Islay, all alone but for my tent, a Ford Transit and a cheerful mongrel dog called Suka. Together, we scoured the wistful autumnal landscape, searching for signs of worms and jelly fish—me in the rocks, and Suka sniffing happily along the shoreline. Layer by layer, I measured the ancient seafloor sediments above the pebble beds at Port Askaig, until we came to a cove filled with a great grey factory. Here, at Caol Ila sat the building in which a fine single Malt was being distilled in fat and gleaming copper vats, each of them the size of a small house.

The air around the cove at Caol Ila was appropriately tainted with the aroma of malted barley, the tang of sherry barrels and the salty smell of bladderwrack.[145] It was, unfortunately, an intoxicating mixture that seemingly went to my head. Not far beyond the distillery, I began to find sediments laid down along a very ancient shoreline. There were all the signs here for lagoons with channels, pebble beds, and mud cracks, much like the creeks of Barbuda. I was ecstatic. Soon, I began to find layers covered with linear and blob-shaped markings, some of which looked very much like the burrows of animals from Barbuda. With my curiosity tweaked, I piled slabs of this rock into my Ford Transit and made the journey back home.

For years I contemplated these strange markings from Caol Ila, never quite sure what to make of them. In fact they were shelved for eighteen years. But each year I would show a colour slide of the blobby structures during my lectures. Little by little, I began to convince myself that the Caol Ila blobs were examples of some very ancient animal burrows. A tentative review was written, and rather too quickly published during 1998. The press seized upon the story, loving 'the world's oldest animal poo' dimension. It was covered on the BBC Today programme and even by the

Six o' Clock TV news. A low point came when a local rag covered it along the lines of 'Oxford Boffin Finds Oldest Poo'. Others opined that I had been sitting on ancient poo hidden in my drawers for nigh on twenty years.

All great fun. But a problem emerged during field work with my colleague John Dewey in the following year. The chain of blobs that had formerly looked to me like animal faecal pellets could perhaps be better explained as dimples caused by microbial mats. Less exciting, of course. But the most boring explanation, by definition, commonly has probability on its side. And that means that a boring explanation is probably a true explanation.

'Wrinkle marks' were only then beginning to be seen for what they were—fossilized microbial mats—in the Precambrian fossil record. They were starting to turn up almost everywhere, too.[146] After a while, it seemed there was far too much time between the Caol Ila blobs—provisionally dated at about 700 Ma—and the first reliable animal traces at about 550 Ma. Not impossible. But my faith in Precambrian animal tracks was severely challenged: I had clearly fallen prey to the fossil Rorschach test. I had found what I was looking for—and not what was really there. Those old Mofaotyof blinkers were about to fall away from my eyes.

To come back to the focus of this chapter, it is possible to see that aspects of the Mofaotyof Principle have been at work in our great antecedents, such as Martin Glaessner. Our aim at that time was to ask: what does the Ediacara biota remind us of? And the answer came back: jellyfish, worms, urchins and shrimps. But the proper question was this: what are these structures really? Recent years have therefore seen a kind of 'open season' declared on Mofaotyof interpretations of the Ediacara biota, with suggested affinities for their biology ranging from ancestral animals to underwater fungi, and from seaweeds to giant deep sea unicells, or even to fish.[147]

This profusion of claimed affinities has proved difficult to resolve, meaning that there is still a lack of clear ancestry for the Cambrian animal phyla, despite considerable progress in technology and thinking over the last century and half. While we may now formalize the questions differently, in terms of cladistics and genetics, chemostratigraphy and geochronology, we still have no clear answers to our long-posed problem: when did the animal groups first appear? The dichotomy between the self-evident appearance of animal fossils at the base of the Cambrian and the uncertain but anticipated prehistory of these groups, means that we still have much to understand about the Ediacara biota.

A major lesson to be learned from the Ediacara biota, then, may be as follows. When we are attempting to answer big questions in science, we must accept that we are bound to make big mistakes. Big Questions = Big Mistakes. That is surely acceptable to us because most big mistakes will be discovered by the process of doubt that is natural to science, given the passage of time. But the converse also carries a warning: those who would only dare to make little mistakes must only focus on little questions. So the risk of 'Big Mistakes' it must be.

Why no jellyfish?

That the fossilized Ediacara biota may contain *no* examples of the major invertebrate animal groups—jellyfish, worms, and all—may all sound somewhat implausible. I am tempted to suggest that a major difficulty here is brought about by the meagre limits of our own, human, imaginations. It could well be that biological reality was very different indeed from the jellyfish world conjured up for us in the 1960s.

In seeking the origins of animal life in the Precambrian, we have been guided by the Great Tree of Life. It sounds wholly logical to be looking for creatures like living sponges, jellyfish, and worms in the Ediacaran period. That is exactly what Lyell's Principle of Uniformity tells us to be looking out for. After all, *the present is the key to the past* is it not? But I think that there are serious exceptions to this rule. And those exceptions are found precisely here, as we start to enter the Precambrian Dark Age. It is here that Lyell's Hunch and its uniformitarian reasoning starts to let us down rather badly.

Our Barbudan 'honeymoon' in Chapter 1 reminded us of the remarkable interdependence between living creatures—between microbes and algae, and between plants and animals. Consider the case of protozoan *Homotrema* that we found growing on the coral reef and extending its pseudopodia out into the current to feed. Being a single-celled protozoan, we might expect it to show 'primitive' features. But almost everything about this little blob is tuned to a world filled with more 'advanced' animals. It sits on top a coral; uses sponge spicules as fishing rods; catches animal zooplankton for food; and secretes a hard shell to stop being eaten by fish. Without such a world, this tiny creature could not exist.

Interdependence with the animal world can also be argued for the cousins of *Homotrema*. In the last chapter, we showed how even simple protozoans like *Platysolenites* took part in the Cambrian explosion, developing a hard outer shell of sand grains glued together. In fact, very similar foraminiferan protozoans can be found living on the deep sea floor today, where they thrive in areas beneath seasonal plankton blooms, capturing the falling bodies of tiny shrimps in their net-like pseudopods. In a

Precambrian world without animal grazers, there would have been little need for its agglutinated shell, and in a world without highly nutritious zooplankton—a free shrimp cocktail—waiting for dinner time on the deep sea floor would have been a forlorn hope. Dinner time may never have arrived.

Today, many protozoans can be found living not so much on the seafloor as within it and beneath it. Before the Cambrian, hunkering down in the sediment would have been rather pointless because, in a world without the circus of worms, the seafloor could well have lacked many of the soil-like components that now make it an attractive abode for life.[148] Instead, the few tens of centimetres just beneath the seafloor—now so habitable for everything from protozoans to worms—would then have been an abode for anaerobic bacteria. These anaerobes would have been about as welcoming to an oxygen-user as Andropov was to Reagan.

What does all this suggest? Organisms of supposedly lowly status—like protozoans, sponges, and corals—are hugely dependent on a world tuned to the presence of higher beings, from worms to molluscs to mammals. This implies that most of these organisms have co-evolved, and could barely exist without each other. As Doug Erwin of Washington has pointed out, a biological species is not an isolated entity—it has an ecological niche which is defined by the presence of other species in an ecosystem. There was no such thing as 'an empty ecological niche' in the Precambrian. Or indeed at any time.

Cnidarians provide a good example of this dilemma. Corals, anemones, and jellyfish are supremely adept at catching little animals higher in the Great Tree of Life, with not only a mouth but an anus. If animals drifting in the sea—called zooplankton—did not really get going until the Tommotian stage of the Cambrian, as Nick Butterfield of Cambridge has recently suggested, then it would barely have made sense for the ancestors of corals and jellyfish to have been covered all over with expensive stinging cells. That would be like opening up a DVD shop in ancient Rome. Without a DVD player and electricity to power it, the Romans would have found little use for a DVD, except perhaps as a grossly overpriced mirror with a hole in exactly the wrong place. Likewise, the ancestors of jellyfish must surely have been designed along different lines before the advent of animals with an anus—the 'bilaterians'. Jellyfish are not primitive beasts. They have co-evolved to prosper in a post-Ediacaran world.

Pores for thought

The same caution can be applied to our concept of sponges in the Ediacaran interval. Because sponges lie at the base of the animal tree, we might expect to find their remains all over the place just before the Cambrian. Provided with tough little spicules of protein, or chalky or glassy materials, their spiky skeletons should have littered the seafloor. Until now, however, no undisputed sponge fossils have yet been found until near the very start of the Cambrian period, and even these are contested.[149] But as soon as the Cambrian begins, we can find little tell-tale signs—cruciform spicules of glassy silica—across a wide area from Iran to China.

Why are sponges so inconspicuous in Ediacaran rocks? Here again, I suggest there is a constructional explanation that relates to their co-evolution with 'higher animals'. Modern sponges can have problems with keeping their delicate pores clean. Without the cleaning service provided by shrimps and brittle stars, the pores will quickly clog in dirty water and the sponge will fade away and die. Sponges are also dependent on a plethora of bacterial particles in the water column. And there is no better way to set up such conditions than to introduce a circus of burrowing worms on to the sea floor. These worms quickly kick up bacterial floccules, both on and within the sediment, which are then entrained by currents in the water column towards the quietly waiting sponges. It may have been possible to find hydraulic conditions that were convivial to sponges before the evolution of bilaterians but the options may have been few. In my view, it can be no coincidence that sponges became both abundant and diverse during the early stages of the Cambrian explosion. By that time, there was a rich bacterial soup for them to feed on, obligingly stirred up from the seafloor by the circus of worms. Before that time, sponges may have been constructed along rather different lines from those we now see.

It can be argued, therefore, that the Ediacaran discs and fronds are unlikely to have been sponges, jellyfish, seapens, or worms. The list of negative attributes is impressive: no mouth, no gut, no anus, no spicules, no porosity, no bilateral symmetry, no organs of filtration, no organs of locomotion, indeed no locomotion to speak of. But enough of 'what they were not'. What were the Ediacaran creatures really, and how on Earth did they live? Our best guess is that they were multicellular forms ancestral to the sponges, jellyfish, and 'sea gooseberries' that still fed largely by absorption, often living well below the zone of sunlight and sometimes well below the sediment surface.

Before the Cambrian, the pattern of life may therefore have worked in ways that were markedly different from those we now see. That could explain why iconic members of the Ediacara biota, such as *Charnia* and *Dickinsonia*, do not resemble more familiar forms of life from later times. These may have been more like corals without stinging cells, or sponges without pores: colonies of cells that were integrated into tissues but not yet into modern body plans.

We must no longer expect to translate our modern world and its biology far backwards in time. The world before the Cambrian was, arguably, more like a distant planet. Duly cautioned, we are now ready to move on to more remote times, to consider not only the world before animals but also the *evolution of thinking about* the world before animals.

CHAPTER SEVEN

REIGN OF THE SNOW QUEEN

A puzzle set in pebbles

Schiehallion is a mountain like no other. Over two hundred years ago, this 'Fay hill of the Caledonians' which rises volcano-like from the flanks of Loch Tay, was used to weigh the Earth, and hence the Sun and planets, for the very first time. Soon after, it became the first to be mapped with contours. Just to make sure, that geological genius John Playfair mapped it and weighed it yet again. This place has a gravity all of its own.

In the summer of 1960, I found myself struggling up the slopes of this Scottish peak. A few steps ahead strode my elder brother, yearning to botanize at the top. I was merely yearning for him to slow down a bit. Together, through bracken and bog, we scrambled into the corrie, a wall of rock shaped like a giant's armchair, scraped out by ice and snow some ten thousand years ago. Monoliths of slate, looking like portals to the hall of the Mountain King, jutted upwards and outwards from the corrie walls. These giant stones had their tender side, though, giving shelter to delicate ferns and lurid green liverworts swaddled in a carpet of brown moss. But carpets of moss were not what we had come to see. Our aim was to find an old flower bed. And there it was: small clusters of alpine gentian and fleabane, trembling in gusts that slid down the corrie walls. This little hideaway, on the flanks of Schiehallion, was a refuge for a carpet of alpine plants that had once spread all the way from Connemara to the China Sea. It was the kind of turf on which the Woolly Mammoth once trod—the Mammoth tundra—a *memento mammori* of a lost world of ice and snow.

Darwin was also much intrigued by the evidence for glaciation in Scotland: 'The ruins of a house burnt by fire do not tell their tale more plainly, than do the mountains of Scotland and Wales, with their scored flanks, polished surfaces, and perched boulders, of the icy streams with which their valleys were lately filled.'[150] And like Darwin, I also found myself drawn back many times in search of an icy enigma, but this was one that Darwin had never seen—the Schiehallion Boulder Bed. I even spent my honeymoon at its feet, from whence my wife's family came. Here, beneath a carpet of alpine flowers can be found the remains of a much older ice age. Not seven thousand years old like the gentian flowers. Not even ten thousand years old, like the corrie walls, but something close to 700 million years old.

This puzzling bed—the Shiehallion Boulder Bed—is little more than a vast jumble of pebbles set within slate. It winds across the landscape like a crazy mountain railway, from the Highlands near Aberdeen, across to Argyll and the island of Islay in the west (see Figure 18). The puzzle bed then dives under the Malin Sea to arrive, dripping wet, on the coast of Donegal. Over the years, I was to follow this strange pebble trackway, plotting its course across both land and chart, and decoding its riddle beside a peaty fire whilst supping a dram—or two—of smoky Islay malt whisky.[151]

By 1975, something like this Schiehallion Boulder Bed was turning up around the world at about the same level—hundreds to thousands of metres below the level of the Ediacara biota and

Figure 18. Realms of the Snow Queen. Map of Western Scotland showing the highlands and islands where the remains of Darwin's Lost World have been sought by geologists over the last two centuries. The Port Askaig boulder bed, which is some 700 million years old, runs from Schiehallion to Islay and thence westward to Ireland. The Torridonian mountains and their 1000 million-year-old fossil lake beds run from Skye towards Smoo Cave. To the west of these are found rocks greatly older than 1000 million years, especially around the Isle of Lewis.

always well below the Cambrian trilobites. From its occurrence in Varanger Fjord in Norway, it became known as the Varanger Boulder Bed. Opinions back then were greatly divided on its deeper meaning. For example, Professor L. J. G. Schemerhorn in Russia had once thought these puzzle beds marked the remains of tectonic landslides that had infected the whole planet. Brian Harland in Cambridge argued for the former presence of icebergs dropping exotic pebbles onto the seafloor, drifting all the way down to tropical waters. George Williams of Adelaide in Australia thought the Earth's axis had wobbled, so that it was the poles that faced the sun, and the equator that froze. Dick Sheldon of Virginia went even further, and suggested the Earth had cooled dramatically when some cosmic collision caused our planet to develop Saturn-like rings, shading the tropics and causing the whole world to catch cold. It was all very puzzling.

The First Cold War

Was the Schiehallion puzzle bed really a monstrous glacial deposit as Brian Harland believed? In 1979, I took a ferry to the Hebridean isle of Islay, to check this out for myself. Islay is not hard and granitic like the Isle of Lewis—which is constructed from some of the oldest and hardest crystalline rocks on the planet—but soft and green, and made of ancient sediments from the Cryogenian period, some 730 to 630 million years ago.[152] Cooked, squeezed, and deformed, these ancient layers of sea bed now twist and turn across the gently rolling landscape. Limey rocks are here clothed in lush green turf peppered with little clusters of blue and yellow flowers, such as harebell and toadflax. On a fine day, it can look like a scene from Tolkien's Middle Earth. Sandy rocks, however, are lower in lime and slower to drain, so support tracts of peat bog with scattered birch. Driving across this landscape, a practised eye could soon infer the kind of rocks that lay beneath.

The road from the Islay ferry to the Boulder Bed crossed the island from west to east, and then wound downhill to a dark little cove called Port Askaig. Gazing east across the Sound of Islay towards the isle of Jura was reminiscent of Tolkien's Frodo staring into the Mountains of Mordor. Jura does indeed look a gaunt and barren place, with two great peaks—the Paps of Jura—standing guard across forbidding plains of rock and moorland. Even the body of water that divides Islay from Jura—the Sound of Islay—has a ghoulish feel to it. Only a few hundred metres across and about ten miles long from east to west, its tidal waters race along like the furies of the night. From the jetty, I watched the old Jura Ferry haul itself carefully across the Sound by pulling itself along an iron chain that stretched across the seabed in a wide arc. On leaving port, the ferry swung rapidly downstream, then slowly hauled its way back up the chain, against the flow, to reach the opposite shore at a place provided with little more than kelp and moss, called Feolin.

The kelp-strewn rocks along the Sound of Islay provide a rich feast for the curious. Not far beyond Port Askaig can be seen our pebbly puzzle bed—here called the Port Askaig Boulder Bed—which is yet another great jumble of pebbles set in sandy slate. Near the Port itself, the pebbles are of mainly of grey and pink granites seemingly floating in a mass of silt. Here and there, I could surmise that these pebbles had long ago fallen vertically, like a depth-charge bomb, through water to land on the seafloor with a plop. Strangely, these pebbles cannot be matched with any granites known in Scotland. More likely, they arrived here from the deep interior of Canada or Sweden. That is exactly what Brian Harland and his students had been arguing—exotic pebbles transported out to sea by ice and then dropped by icebergs on the seafloor. A glacial interpretation for the Schiehallion—and hence Port Askaig—pebble bed therefore seemed sound.

But there was still a nagging problem that emerged more clearly when I travelled down to the remote Mull of Oa, in the south of the island. Here the glacial puzzle bed lies atop a bed of Islay

Limestone. This chalky rock called to my mind the Barbudan lagoons and could be taken to signify the former presence of tropical waters. How could such icy conditions ever have taken hold in the tropics? During the last Ice Age, for example, the summer snowline seldom reached latitudes lower than 50 degrees north—places like London and New York. But in the Cryogenian, ice seemingly reached right down to the equator itself. If that was true, it was highly paradoxical.

All manner of ideas were being put forward to explain this seeming paradox, of ice sheets reaching right down to the equator. And these explanations were starting to demand a world that went far beyond the realms of Darwin and Lyell. A world gone mad. Only Bravehearts would dare to venture on to that turf.

It was Joe Kirschvink of the California Institute of Technology who first rattled the sabre. Joe was one of many geologists working on the preservation of magnetic fields in ancient rocks—affectionately called 'palaeomag' and rather less affectionately called 'palaeomagic'. Over the years, Joe has offered a smorgasbord of ideas for scientists to feast upon. Some of these are quite homely, like magnetic tracking systems in homing pigeons. Others seem more wild and wonderful, like an axis of rotation for our planet that has wandered across the surface of the globe like a lost child—the so-called True Polar Wandering. And one idea was just downright mad—the Snowball Earth Hypothesis. It is this last, crazy, idea that really interests us here. 'Snowball Earth' is a hypothesis that paints a picture of a world in which the ice caps expanded from the poles right down to the equator.[153] They grew to such an extent that the oceans were entirely covered in a continuous sheet of sea ice, several kilometres thick. Landmasses of the time may have been very cold and dry—with average temperatures down to minus 50 degrees Celsius or so—with almost no land ice at all.

Meanwhile, seafloors were insulated from this cold by thick sea ice, with life on the bottom either as snug as an Inuit in an igloo, or starving to death.

Joe Kirschvink's Snowball hypothesis could well have languished in respectable obscurity had it not been for two Harvard scientists, Paul Hoffman and Dan Schrag, who leapt to its aid, gathering information and broadcasting every new twist with enormous energy. Using emerging evidence from the chemistry of the rocks, they were able to point out that dramatic swings had taken place in the carbon and climate cycles at this time. These, they showed, had seemingly swung between something like the present climates of Mars and Venus. Very cold and very dry during the glacials—when the cycles of water and carbon were almost shut down for tens of millions of years, and very warm and humid during the interglacials. The planet seems to have swung into reverse—from Mars to Venus—in just a few thousand years. It was a truly hellish vision. But was it true?

Learning the Dark Arts

For many people around the world, Snowball Earth was little more than a horror story, a kind of geo-entertainment. That was fair enough when it was thought to have rather scant implications for our daily lives. But in one place in the world, it really mattered. Not in Harvard, where the Snowball prophets preached, nor yet in Washington, where the Global Warmers prayed. It was in Oman, where the oil was in danger of running out.

Mineral oil is born from remains of algal blooms that floated on the surface of the sea, during sunny days long ago. Not any old sea, though—the seafloor conditions had to be a touch on the toxic side of 'smelly', too, if the oil was to end up inside the rock rather than inside the bellies of worms. Millions of years of seafloor sinking and sediment burial have then dragged these algal oils downward, so that they are now found sleeping in beds that lie really deep, ideally some five to ten kilometres down. Here in the depths of the Earth's crust, the heat radiating upwards from the Earth's mantle has gently cooked the oil and turned it into a useful kind of broth. Finding this black broth, though, is like looking for a black cat in the darkness of the night. Or even more like looking for an enemy submarine in the depths of the ocean. Our oily denizen of the deep has to be made visible, using a kind of echo-location called seismic surveying. Tracing the oil back to its den is the really tricky bit. But getting it 'out of bed' is pretty difficult too—worse, even, than getting my students out of bed for a nine-o'clock lecture. And getting our black broth back up to the surface is no easy matter either. All of this needs something that we may call the Dark Arts of Geology.

Oil companies specialize in these Dark Arts—men and women who are highly skilled at sniffing the gasoline back to its lair. Drilling for oil is expensive, costing millions of dollars a day. A dry oil well is not only a bit of a disappointment. Like an empty gin bottle, it is a potential step on the road to ruin. Oily secret agents therefore live their lives, like stockbrokers, in a state of nervous anticipation which only numerous and exotic vacations can ever hope to repair.

Over in neighbouring Saudi Arabia, the risks of the oil fields running dry still seemed rather low. Much the same was true for oil fields in Iran, just across the Arabian Gulf. But for Oman, a cruel trick of history has meant that it does not share easily in this bonanza. The Saudi fortune came mainly from oil that formed during the age of dinosaurs, the Mesozoic, when oily algal blooms reached a kind of peak. Part of the problem for Oman was that rocks of this age had been folded into mountain chains along the Arabian Gulf. These mountains—the Jebel el Akdar—are certainly very scenic, with their lofty peaks and palm tree groves, as we shall see. But their oily rocks are too near the surface to provide the cooking and pressure needed to make them useful. To the south, however, lies a great desert—the edge of the Rubh al Kali—the dreaded Empty Quarter of Arabia—a mournful place of canyons, crags, and sand, stretching across to the horizon. There is barely a whiff of oil at the surface here. But deep down, the oil company experts had detected enough clues, by the mid 1990s, to form a serious hunch and to hatch a cunning plan.

The national oil company of Oman[154] suspected that there was black gold at surprisingly ancient levels in the rocks—from about the level of the Snowball Earth and up to the Cambrian explosion. Millions of barrels of it. What they needed next was a kind of secret map to show the way the world worked at that time. They needed a stratigraphy. If they could get that secret information out of an informant, then Oman might be able to strike it rich more rapidly. And that was how I accepted the bait and became an informant, to help reconstruct a secret sketch of the world that lies beneath Oman.

Little clouds of chalk

Soon after arrival in Muscat, my Oxford colleague Philip Allen and myself were whisked off to enter the great enclosure of the oil company, sitting beside the bright and breezy Arabian Sea. Brilliant red bougainvillea flowers bedecked the bright white walls of its Core Shed. Once inside the cool of this building, we were presented with an awe-inspiring sight. The Core Shed is like the inner sanctum of a geological temple. It enshrines those chapters and pages rescued from the geological bible—many kilometres of core, drilled from rocks deep down and now brought to the surface, to be stored in trays, neatly stacked in columns reaching 50 feet high.

It was here in the Core Shed that we saw, for the first time, the earliest possible evidence for animals with shells. These strange structures had been brought to the surface from boreholes far to the south, where they had been found at the level of the ancient oils we were looking for, just below the start of the Cambrian era and some 542 million years old. The mottled cylinders of marble that made up the rock core contained, here and there, sections through at least two kinds of tubular shell. The first was a rather ragged kind of tube, formed from a stack of delicate collars, named *Cloudina* by African geologist Gerard Germs, after the American geologist Preston Cloud.[155] We had no idea what kind of creature lived inside these tubes. Most palaeontologists favoured the idea it was some kind of worm-like animal, like a modern feather-duster worm, though it could have been made by some simpler creature like a protozoan or a coral. Either way, it does seem to have succeeded in secreting the first known calcareous shell. The second fossil was shaped like a conjoined nut-and-bolt, and is called *Namacalathus*.[156] Appropriately, the hollow rounded 'nut' has about six facets around the flanks, each provided with a rounded opening (see Figure 19). Although compared with sponges and corals, this strange object has rather little in common with those fossils we have met at the start of the Cambrian. Another possibility, therefore, is that *Namacalathus* was a *Charnia*-like fossil, in which the 'nut' served as a basal attachment for a frond that was otherwise not preserved (see Figure 19). In this view, its strange appearance arose in part from the accidental calcification of a soft-bodied Ediacaran organism on the seafloor, a bit like the formation of fur on the inside of a kettle, or even more

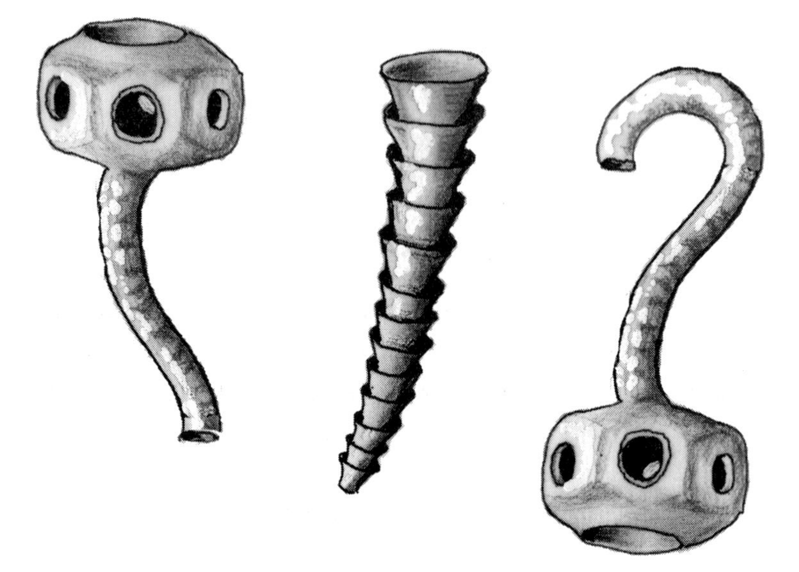

Figure 19. Little clouds of chalk. The Ediacaran shelly fossils from Oman and Namibia are from 549 to 542 million years old and include *Cloudina* (at centre) and *Namacalathus* (on either side), here reconstructed by the author. These fossils are usually about 10–40 mm in length. *Cloudina* was probably made by a worm-like organism. *Namacalathus* is altogether more complex with its 'cup' at the top (at left) or part of a basal holdfast at the bottom (more speculatively, at right).

like the crust of chalk called 'tufa' that can form around a piece of moss or a can of coke, where lime-laden waters spring to the surface.

Into the Valley of Dates

Soon after paying our respects to the inner sanctum of the Core Shed, we were whisked off to the Jebel-Al-Akhdar mountains by our Omani guide, Salim Al Maskery. At the airport, Salim looked magnificent, dressed in turban, dishdasha gown, and khanjar silver dagger, but he wisely donned boots and jeans in readiness for the field. The Omani mountains were reasonably close, about two hours of driving—most of it nearly suicidal, of course—from the centre of Muscat. Oil companies in Oman were, at that time, suffering about seven staff fatalities a year due to accidents, mainly on the roads, and their safety protocols were accordingly strict.

Once we were over the shock of the traffic and its speed, the journey into the hills proved a tonic. One might be forgiven for thinking that the scenery of the Jebel-Al-Akhdar mountains was brought together as the set for a 'Chelsea Tractor' (SUV) advertisement in, say, a glossy airline magazine. The stony track from the highway wound through deep gorges filled with massive boulders, sometimes along the floor of the valley and sometimes along a twisting 'path of death', cut right into the mountainside itself. Here and there, we would come across little oases, where farmers were herding goats, or growing groves of dates and figs. Driving ever deeper into the heart of the mountains, we finally arrived at our destination, in Wadi Bani Kharus. The hot Toyota engine at last fell silent, leaving a cloud of dust to roll in from the road behind. The angry roar of the motor was suddenly replaced by the gentle sound of tinkling goat bells.

Looking at the map, we could see that the rocks in this wadi were among the lowest—and hence the oldest—rocks exposed in the whole of the Jebel. A quick scramble across the rust-brown crags, already warming in the morning sun, quickly showed us a great jumble of pebbles set in slate. Here, in this exotic place, I felt a twinge of *déjà vu*. That old Pebbly Puzzle Bed had come back to haunt me once again. But this Omani Puzzle Bed was to prove a bit special. It concealed a hand of trumps.

That evening, under a spectacular star-studded sky, we gathered around a camp fire built up from fragrant, resin-scented logs, and talked far into the night about dates from Arabia. We were not thinking here about oases with groves of date palms. It was dates of another kind that caught our attention, dates to help reconstruct our plan: the dates of rocks.

Dates were, and are, really important to oil sniffers because 'dates mean rates'. And rates play a key part in any search for patterns in the distant past. Over the following weeks, therefore, we explored the hills and valleys of the Jebel-Al-Akhdar with great attention, looking for rocks that could yield valuable dates. The scenery in which we searched often reminded me of the Valley of the Kings in Egypt. Indeed, our search also called to mind the case of Howard Carter, searching for the tomb of Tutankhamun. In 1922, Carter knew that one king had not yet been found in the valley. His name had even been erased from the official King List. But he knew that King Tut existed because of finds in the valley of things with his logo on—things resembling souvenir mugs from the Bronze Age. Carter's ancient King List gave the lengths of all the other reigns, from which it could be estimated that King Tut had reigned some 1300 years ago. But for us, working in Darwin's Lost World, there was nothing like Carter's King List—no clear evolutionary succession of fossils. Nor was it known how long the reign of either 'The Snow Queen', or 'King Ediacara' lasted. To put that list together, we were hoping to find rocks that contained fossils. Or rocks with marker beds that could be traced to other fossils. Or, best of all, we were looking for rocks with volcanic ash beds.

Ash beds hold a special place for geologists. That is because volcanic eruptions contain pretty little gemstones, called zircons, whose not so pretty little 'radioactive clocks' are reset like an egg-timer at the start of each big eruption. Zircons are tough little crystals and keep the time with amazing precision, come hell or high water.

Snowball hell and Cambrian high water were exactly what we were hoping to date, so our plan was to look out for ash beds that might contain zircons. And within a week, we were to hit the jackpot—in the Gubrah Bowl.

The Gubrah Bowl is a great depression in the middle of the Jebel-Al-Akhdar, shaped like a vast Roman amphitheatre. Eons of erosion have sculpted the landscape here into a ring-like rim of tough younger rocks—of mainly Mesozoic age—leaving a central bowl of older, softer rocks. It was these older rocks that we had come to see. In the middle of the dusty Gubrah Bowl, we found a hill emerging out of the thick alluvium—it was yet another glacial puzzle bed. But this puzzle bed was seemingly rather different from all the others. Through it ran a thick layer of volcanic ash. Ash beds, as we know, mean dates, and dates mean rates. In no time at all, therefore, we had it measured, photographed, sampled, picked, and sent out for zircon dating. In a few months, a series of readings from the zircons came back from Bob Tucker in Michigan. This ash in the Gubrah puzzle bed was about 716 million years old, much more ancient than we ever expected. Assuming the dating of the ash was reliable, which later proved to be the case, it showed exactly when the global reign of ice and snow might have begun.

A single date on its own does not offer much help, of course. With later evidence from Oman and beyond, though, researchers were able put together a list of dramatic events like the one below. This list is laid out like an ancient Egyptian King List, with the oldest events placed at the top.

716 First reign of the Snow Queen, Gubrah Glacial
640 Second reign of the Snow Queen, Sturtian Glacial
630 Third reign of the Snow Queen, Fiq (Marinoan) Glacial

580 Last reign of the Snow Queen, Gaskiers Glacial
575 Start of the reign of the Ediacara biota.
549 Appearance of the shelly fossil *Cloudina*
542 End of the reign of the Ediacara biota. Entry of the Circus of Worms
530 Start of the Tommotian Commotion
525 Preservation of the Chengjiang biota

This 'King List' refers to millions of years before the present. Such a list really matters for several reasons. First, to the Oman oil company, it meant that the history of the oil-bearing rocks could now be predicted, using both computers and clever sets of sums. But for us and for Darwin's Lost World it matters even more. Dates help to give both shape and context to the prelude to the Cambrian explosion. Not least, these dates were starting to reveal at least four glaciations spread over a huge span of time, not only one or two, as at first had been thought.

Wadi Sahtan and the Coffin of Doom

Among the set of questions that still intrigued the world of geology, however, was this one: *what was the extent of sea ice during each Snowball glaciation?* If it *was* really continuous from pole to pole, then the consequences for the biosphere would surely have been catastrophic.

The prophets of Snowball were inclined to believe that, during each Snowball glaciation, the kilometres-thick layers of sea-ice formed a kind of lid across the oceans. Shut away from both sunlight and solar warmth, photosynthesis could have dwindled to a few kinds of microbe living in suspended animation within the ice sheet. Conditions could have been even more dire for bottom-dwelling animals, cut off from their supply of oxygen. One kind of refuge and a source of nourishment, perhaps, was those hot springs that tend to well-up along the mid-ocean ridges, exactly where the Black Smoker ecosystems now thrive. Even so, the long reign of ice and snow would have been rather like being buried alive in a coffin. Only the Goths of the biosphere could ever have survived such entombment.

Wadi Sahtan—the Valley of Satan—is an appropriately gaunt and claustrophobic valley in the Jebel-Al-Akhdar, hemmed in by high and devilish-looking hills. On a cloudy day, it can look a bit like the setting for a painting by Breugel the Elder of the 'Last Judgement'. On the north side of Wadi Sahtan is a dry waterfall that climbs, ever more steeply, towards the top of the cliff. And here, a mountain stream has cut through stacked layers of what we decided to call the Fiq Boulder Bed. This demonic gorge was a perfect place to test the truth of our nightmarish vision—of life being buried alive in its own coffin.

Our doctoral student, Jon Leather, led Philip and myself up the waterfall, so that we could follow the layers, bed by bed. Sure enough, we stumbled upon jumbled layers of pebbles set in slate, brought about by ancient icebergs that had dropped huge exotic stones down on to the seafloor. But every now and then there came a marked change in the colour of the cliff, owing to ancient beds of sand with ripple marks. Not any old kind of ripple marks, either—these were the ripples left by waves that had crossed an open body of water. Clearly, the sea ice in this place cannot have been like a coffin lid. Instead, the ice had broken up, from time to time, to leave open bodies of water that lasted for many thousands or even millions of years. Similar intervals of open water were found at higher levels within the hill of Wadi Sahtan. In fact, the lid of ice had come and gone as many as seven times.

On seeing this, it felt as though we could breathe fresh air again. Jon Leather had helped to put a silver nail into the Snowball coffin. The picture of early life being buried alive in its coffin seemed as good as dead, and the Snowball hypothesis was starting to look a bit ragged as well. Wadi Sahtan was showing us that the surface of the sea must often have been like a modern glacial ocean, such as the modern North West Passage in the Arctic—a mix of ice sheets, icebergs, water and slush. It was still a reign of snow and ice. But it was a Slushball rather than a Snowball world.

It was then that we noticed something rather odd in the story book of rocks that lay open near the waterfall. Sure enough, every time the ancient sea ice had melted away, a bed of jumbled pebbles gave way to beds of rippled sand. But there was more. The seafloor also became covered with a brownish coloured chalky deposit, called dolomite. Following these chalky layers up the cliff, we could also see how the reign of the Snow Queen had literally come to a sticky end. The last glacial beds of pebbles in slate were abruptly succeeded by a huge pile of dolomite, ten or so metres thick. Everywhere we looked in Oman, we found such a layer of dolomite, capping the last of the glacial pebble beds.

The post-glacial burp

What are we to make of the periods between the glacials? What happened then? It was here that these dolomites above the glacials—the so-called cap carbonates—started to tell their own very strange story. Every major Snowball glaciation was, we soon learned, terminated by such a cap carbonate in many parts of the world. To test this, we were later able to follow these caps into Canada, across into Scotland, down into Namibia, up into India, over the Himalayas into China, and then down into Australia.

Figure 20. A 'golden spike' has been placed at the base of the new Ediacaran System in the Flinders Ranges of South Australia. It lies in the bottom few centimetres of the cap carbonate, and just above the Marinoan glacial deposits—of Snowball Earth fame. As such, it marks an episode of seemingly rapid melting of near-global ice caps. The circular holes in the rock were left behind by cores taken to test, and indeed to confirm, that glacial deposition took place near to the equator at this time, some 630 million years ago. Image taken by Jon Antcliffe.

Indeed, so dramatic and widespread were these marker beds that one of them—a few centimetres above the base of the Marinoan cap—has been taken to define the base of the new Ediacaran System across the globe (see Figure 20).[157] Arguably, it is the biggest, single, planet-wide marker bed. Only the famous meteorite bed at the end of the Cretaceous comes anywhere near, though that is usually only a few centimetres thick. Cap dolomites like this can be tens to several hundred metres thick. It is the chemical signals inside these cap carbonates, though, that makes them a bit spooky. They contain ratios of carbon isotopes that resemble a kind of giant 'burp'. To understand this we will need to take a closer look at the carbon isotopes of a belch.

We all know that life on Earth is carbon based. Many will also have heard of carbon-14 dating. That is enough to get us started here. Carbon-14 is a radioactive isotope of carbon, which is why, when it breaks down naturally into nitrogen-14 at a characteristic rate, it is used as a 'test of time' for archaeologists wanting to date old bones. Carbon-12 is a relatively skinny little isotope when compared with carbon-14. It is therefore called the light isotope, because it weighs in at less. But there is also another 'stable isotope' that weighs in at a slightly chunkier 13. It is the ratio between the bantam 12 and the middling 13 that we need to focus on here.

Take a deep breath. And now breathe out again. Curiously, the air you just expelled is richer in carbon-12 than the air you breathed in. In fact we can be fairly sure that it was enriched in the lighter isotope by a factor of about 28 units in a thousand. Your dinner is partly to blame for this because the carbon in the food on your plate was either made by plants, or made by animals that ate plants. Or if you happen to be a top predator—a well-educated vampire perhaps—it was made by animals that ate animals that ate plants. Either way, it's the green plants that cause this *fractionation* of carbon. Plants show a bias towards lighter carbon-12 because its lesser mass allows it to be more easily incorporated into biological materials than carbon-13. As the saying goes, 'life is lazy'. And this sign of laziness gets into coal, oil, and most other rocks. The carbon in the gases chugging out of the exhaust pipe of your car has a carbon isotope ratio biased towards carbon-12 as well.

But now imagine that you are sniffing a sample of the gas coming out of your nearest volcano—say, Etna or Old Smokey.

This gas has much more carbon-13 than does your breath. That is because the realm of plant life—oil and coal and chalk—barely reaches down to the depth of the magma chamber of the volcano, many kilometres below. The carbon in gases exhaled from volcanoes is only enriched in the lighter isotope by a rather modest factor of about 5 units in a 1000.

What Paul Hoffman and his colleagues at Harvard were finding was that the carbon isotopes in the cap carbonates were formed from something like a burp. One idea is that they formed from pent-up carbon gases, previously trapped not only in sediments but in seawater beneath the ice, that were rapidly vented to the surface. As these gases met up with warmer surface waters, the oceans then turned white with chalky matter, which settled down to form the layers of cap carbonate.

Flatulence may be socially embarrassing, but doctors still regard it as a good thing for digestion. Thus it is, then, for the planet at large. Oceanic dyspepsia during the age of ice was eventually relieved by a good old-fashioned bout of flatulence. Flatulence is not without its drawbacks, though. Most notably, it can lead to global warming. For example, about 14 per cent of modern global methane in the atmosphere comes from the guts of cows and other farm animals. In a similar way, bouts of global flatulence—recorded by the cap carbonates—are thought to have precipitated the dramatic shift from an icehouse world to a greenhouse world, possibly in just a few thousand years. A jump from a climate like that of Mars to that of Venus would be taking this story too far, but it was probably one of the most hellish series of shifts that the Earth has ever suffered. The reign of the Snow Queen seemingly came to an end when this pent-up methane and carbon dioxide gas, some of it from volcanic outgassing, finally burst through the ice.

The 'Creosote effect'

Did anything else happen between the glacials? It was with this question that we began to run out of data. Fossils that came from these interglacial phases are still surprisingly poor and rare. Carbon isotopes from these times do, however, tell us something rather suggestive. They show how the oceans recovered from the giant burps, only to overshoot the mark and achieve some of the most extreme values ever recorded. It may be here that a kind of 'Creosote effect' came into its own.[158] As the water cycle got started again—rivers began to flow, and mountains began to erode—the seas may have been fed with nutrients to the point of bursting. The bottom layers of the oceans then became a bit like a cess pit—stagnant with the decay of unwanted food—to the point of anoxia. Without zooplankton or animals to sweep up after this mess, parts of the ocean became stagnant to within a few tens of metres of the surface. Huge amounts of carbon seem to have been sucked down into these foetid lower layers in this way, where they were buried almost forever in the mud. That was a profligate waste of vital resources. It helped to produce the oil in the rocks, of course. But it may also have led to a cruel retaliation—the eventual return of the Reign of the Snow Queen. That is because carbon dioxide is one of the key greenhouse gases, and its removal into the cess pit of the rock record could—in theory at least—have led back towards a cooling, and maybe to an extreme cooling, of the atmosphere. Or so the story goes.

So let us take another look at our 'King List'. Only this time, we shall set it out like the layers in rock—with the oldest events at the bottom.

542 End of the Reign of the Ediacara biota
575 Start of the Reign of the Ediacara biota

580 Last Reign of the Snow Queen, Gaskiers Glacial
Inter-regnum
625 Beginning of the new Ediacaran period
630 Third Reign of the Snow Queen, Marinoan Glacial
Inter-regnum
640 Second Reign of the Snow Queen, Sturtian Glacial
Inter-regnum
716 First Reign of the Snow Queen, Gubrah Glacial

As we can see, the reign of the Ediacara biota lasted for a period of about 40 million years or so. But it had been preceded by a long period of turmoil, the so-called Cryogenian period, during which the Earth was buffeted from one kind of Hell to another. From this long list, then, an equally long series of nightmares was starting to emerge.

Did Snowballs induce a birth?

We now come to a very big and difficult question. Could the snowball—or even the slushball—glaciations have helped to bring about, in some way, the Ediacaran and Cambrian evolutionary 'explosions' that followed? There is certainly a very strange coincidence that needs to be explored here—Earth's most severe glaciations came to an end about 580 million years ago, only to be followed in quick succession by the appearance of the first large animal-like remains in Newfoundland, about 575 million years ago. As you have probably noticed, this is a variation on Daly's Ploy: that big evolutionary changes are driven by big environmental changes in the planetary surface—in this case relating to climate as well as to seawater chemistry.

This debate, and that for Daly's Ploy, reflects our constant search for a satisfying relationship between cause and effect. We are always looking for simple explanations for things: why was there a traffic jam on the way to work? Why did your mother catch that cold? And why did your father fail to get promoted? Each of these example comes from a complex system, in which the relations between cause and effect are unpredictable, and so they seem worrying to us. There is, therefore, a natural but wholly futile human tendency to look for simple explanations where none may exist. The daily newspapers are full of examples: who is to blame for the collapse in the dollar? Why did Manchester United lose the Cup? Even my own subject of Earth Sciences has been susceptible to this kind of tabloid thinking.

Nowhere is it better displayed than in the burgeoning field of Climate Change and Global Warming, where the chain of logic typically runs as follows: carbon dioxide is directly associated with global warming in the past. Carbon dioxide levels are rising today, and so are global temperature levels. That means that carbon dioxide is the *cause* of global warming. And that means that its *effect* is the global warming we now experience. Hence, carbon dioxide levels must be reduced if we are to reduce global warming!

All this may be true—and I am as concerned as the next person about this issue. But there is a potential for error in the chain of logic here. It is always necessary to prove whether rising carbon dioxide levels are *the cause or the consequence* of global warming. Had the latter been true, then our chain of logic would collapse. And there is a further problem with this line of logic as well. It is widely assumed that the rise in carbon dioxide and the rise in global warming will behave in a predictable way—that is to say, in a mathematically linear way. But there is reason to suspect that it will behave in a rather unpredictable way, more like a chaotic system.[159] If that were so, it would mean that the outcome of remedial policies is altogether uncertain. And most especially, it means that sprinkling the surface of the oceans with iron from rusty old automobiles—to increase the uptake of carbon dioxide by the plankton—is dubious at best, and could even be downright dangerous. The remedy might be worse than the symptom.

So to return to our thinking about early animal evolution, how does Daly's Ploy typically run for the Snowball Earth? As we have seen, for the most part, the argument usually takes something like the following hourglass shape:

Snowball glaciations—down to the equator—dramatically reduced the habitable area.
The main refuges for life therefore lay far below the ice, or around submarine
hot springs, where geothermal energy could replace the loss of energy
that had once been supplied from the sun via photosynthesis.
Smaller population meant that an evolutionary
bottleneck was created, through which
all of life was obliged to pass.
With vastly reduced
populations, the
genetic diversity
was also
greatly
reduced.
Later deglaciation meant
that some lucky individuals would
see their genes spread throughout the population with time.
Even oddball mutations would find a niche. Oddballs like the Ediacara
biota, for example. Once the snowball earth came to an end, these Lucky Jims
of the early biosphere would multiply and perhaps eventually come to dominate...

This scenario—of an evolutionary bottleneck—is close to the way in which both Darwin's finches, and the Galapagos tortoises are thought to have to formed new species on their respective islands.

It is the Founder Effect. And that is how many new species and varieties are thought to have arisen, even today.

All this is fine and dandy, except for one little thing. As with carbon dioxide, methane, and Climate Change, we may have allowed ourselves to be led into error with regard to the identification of 'cause' and 'effect'. Extreme climate change is assumed to have 'caused' an 'effect' which, in this case, is the evolution of our animal ancestors. But that is perhaps rather questionable as logic. Our animal ancestors could equally well have been the *cause* of extreme climate change during the Snowball Earth. And there are some good reasons for thinking that may have been the case.

Consider the following lines of thought, for example. Simple prokaryote cells, like those of bacteria are very small indeed (mostly about one-thousandth of a millimetre—a micron—across). Eukaryote cells are much larger (from tens to hundreds of microns across) because they have more things inside them, handy gadgets like the mitochondria and the nucleus. Because these eukaryote cells are bigger than prokaryote cells, they unavoidably have a lower surface area relative to their volume.[160]

Now this is where it gets interesting. Given their lower surface area, these larger eukaryote cells will likely decompose more slowly than smaller bacterial cells.[161] That means that eukaryote cells may more easily be buried in the rock record as hydrocarbons, including oil and gas.[162] In soft sediments on the oceanic seafloor, this inferred greater burial of organic matter could therefore have led towards a greater relative removal of gaseous carbon dioxide from the atmosphere.

One potential fallout from the evolution of multicellularity may therefore have been a greater potential for cooling of the planetary surface—no bad thing when our sun was daily growing ever hotter.

So much for larger cells. But there is also the effect brought about by colonies of larger cells. Such colonies may have been better than single cells at resisting, say, the nibbling effects of grazers. Multicellular organisms could also provide their colonies with increasing levels of specialization for such things as reproductive cells and basal attachment structures. So here again we come face to face with the fact that multicellular colonies are likely to be physically bigger than most single cells. They will, likewise, tend to have a lower relative surface area and they will tend to decompose rather more slowly, other factors being equal. And they may, again, more easily be buried as carbonaceous matter in the seafloor sediment. So yet again, we meet up with a feedback loop in which the evolution of multicellular eukaryote cells may have encouraged the greater removal of carbon from the atmosphere and led towards a potential for further cooling.

In this way, conditions that—in the world of little cells—would have merely caused a slight cooling of the atmosphere were now able—in a world filling with big cells—to cause ice ages that reached right down to the equatorial regions. Looking through the lens of life in this back-to-front way, evolution of the biosphere could have been the *cause*—through some kind of positive feedback—of the Snowball Earth.[163]

So is there any evidence for these big cells and their even bigger multicellular colonies in the world before the Snowball Earth? To search for their beginnings, we now need to pay a visit to some even older mountains.

CHAPTER EIGHT

THROUGH A LENS, DARKLY

Quinaig's Pyramid

I felt obliged to stop the vehicle owing to a growing sense of *déjà vu*. My students and I had been motoring along the road from Durness to Loch Assynt in north-west Scotland, snaking up a mountain pass through the famous Pipe Rock—a pile of fossilized seaside some forty furlongs thick. This Pipe Rock was a buff to flesh-pink sandstone packed here and there with worm tracks shaped like the pipes of Pan. But it was not the pipes that got me. It was the panorama of rock and bog, lying alongside the road from Kylesku to Inchnadamph.

On this very hill, more than thirty years ago, I had led my first training camp in the skills of geological map-making. As a young university lecturer, I had been advised that this landscape could conceal the answers to Darwin's Dilemma, perhaps lying hidden deep beneath its mantle of turf and snow. And key to it all, to the right of the road, rose Quinaig in all its majesty. This mountain rises above Loch Assynt in much the same lofty way as Khufu's Pyramid rises above the west bank of the Nile (see Figure 21). Except, of course, that Carn Quinaig is about ten thousand times as big and 250 thousand times as old.

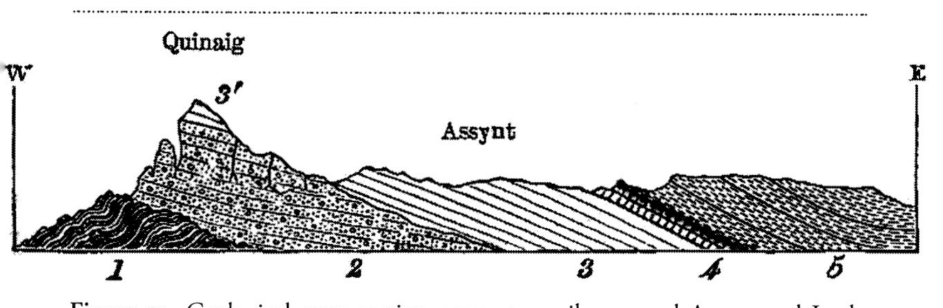

Figure 21. Geological cross section across ten miles around Assynt and Loch Torridon, adapted from the drawings published by Sir Charles Lyell and referred to in the text. His key to the rock layers is here adapted as follows. (1) Crystalline rocks (gneisses and schists) of the Lewisian complex, now known to be from 3000 to 1000 million years old. (2) Bedded red Torridon sandstones of ancient rivers and lake beds here gently inclined towards the east, from 1300 to 800 million years old. (3) Bedded grey quartzite sandstones of the Pipe Rock, Fucoid Beds, and Salterella Grit, of earliest Cambrian age, here more steeply dipping towards the east, and some 540 to 500 million years old (3' marks an outlier of these rocks on top of Quinaig). (4) A zone of major faulting called the Moine Thrust, which took place roughly 400 million years ago. It lies mainly within and above the Cambrian carbonates of the Durness Group. (5) Ancient crystalline rocks of the Moines have here been pushed up from deep inside the crust and thrust on top of the younger Cambrian; they are some 1000-700 million years old.

Sir Charles Lyell seemingly felt in awe of this mountain as well: 'I had an opportunity in the autumn of 1869 of verifying the splendid section [given in his figure 82] by climbing in a few hours from the banks of Loch Assynt to the summit of the mountain called Queenaig, 2673 feet high.'[164] Now Lyell was a carefully logical thinker. And he had a good deal more than a single mystery on his hands back then. On climbing to the top of Quinaig and scanning the horizon he was, of course, confounded by the dilemma of Darwin's Lost World. But the view from the top of Quinaig confronted him, and it later confronted me, with three other puzzles—some of the biggest riddles in the whole of

Earth history. Had Charles Lyell kept a pocket diary, he might have licked the tip of a pencil and jotted down a list that looked something like this:

Consider:

1. *A mountain that stood on its head*
2. *An ocean that disappeared*
3. *A rock that swallowed time*

We now need to grapple with these very same riddles in order to read the lie of the land around Quinaig. Once those riddles have been solved, and their lessons learnt, we will be better equipped to guess the name of the game and reach the end of our quest.

A mountain that stood on its head

Back in 1975, the road below us was much slower-going and more desolate, just a single track with passing places, winding down towards the ferry at Kylesku. In those days, the latter resembled a lost parcel of Greenland—a modest cluster of cottages with neither gas nor mains electricity. The chugging of a generator told me when it was time for a breakfast of tea and porridge liberally sprinkled with fresh butter. Thus fortified, I would drive through flurries of snow to the inn at Inchnadamph, to rally my students through the rocks.

Looking north from the door of the inn, we could see the Highlands. These rise above the 'lowlands' as would a stone wall above a road. The lowland 'road' is made of Cambrian sediments with fossils. And the highland 'wall' is made of tough, old crystalline rocks—the acid heart of the Scottish Highlands—and more properly called the Moine granulites, or 'the Moines' for short.

Now, the Moines present geologists with a bit of a puzzle. In fact, they present a very big puzzle. At the time of Darwin's writing in 1859, it was widely agreed that such crystalline rocks had been cooked and squeezed at high temperatures and pressures. But how did such highly cooked rocks get to lie above rocks that yield well-preserved trilobites which had never been so cooked or squeezed?[165] It was like finding a layer of frozen ice cream below the pastry of a piping hot pie—a Baked Alaska made in rock. That conundrum was not at all easy to explain. James Nicol of Aberdeen came close in 1858. But it was not until 1907 that the convincing solution came to hand, from the heads and hammers of Ben Peach and John Horne. This dynamic duo produced a series of reports that culminated in a big fat 'Geological Memoir' on the North-west Highlands of Scotland.[166]

The rules laid out in the 'Memoir' of Peach and Horne were simple enough to follow: map the country at all scales from kilometres to microns. Then collect and analyse the rocks and fossils using the latest techniques. And then test your ideas with geometry and chemistry. Following these simple homely rules, Peach and Horne helped to reveal that the rocks of the Moines were not quartz sandstones—as old Hugh Miller had thought, back in about 1841. Nor were they squashed-up Silurian rocks—as Sir Roderick Murchison had thought in about 1858.[167] They were, instead, much older portions of crust that had been squeezed up to the surface from deep down to form those mountain ranges that today include the Scottish Highlands. These Moine granulites of the Highlands had then been pushed outward to the west where they became rolled over and even turned upside down, rather like the folds on a rumpled table cloth. This rumpling was accompanied by massive dislocations called thrusts, so that the Moines came to lie on top of younger Cambrian sediments with trilobites, which can be seen at that locality near

Ullapool called Knockan Cliff (see Figure 21, near 4). That is why the contact of the crystalline Moines with the underlying Cambrian rocks here shows all the signs of crushing and alteration.

An important lesson had therefore to be learned in these hills—that mountain belts can be turned upside-down, and they can be turned inside-out. Here, from Inchnadamph in the north, down to near Perth in the south, was found a pile of rocks that had been turned over and 'stood on its head' by the restless movements of the Earth's surface.

An ocean that disappeared

Standing again in front of the inn at Inchnadamph, we could also see, beneath the crystalline Moines, a cascade of white and grey rocks. These are arranged in a series of steps looking like a giant's stairway (see Figure 21, near 5). As with the Moines, they can be traced along the western margin of Scotland, from Skye in the south up to Durness on the northern coast. But unlike the impoverished acid soils of the Moines, these limestone soils sport a rich carpet of calciphilic flowers amid bright green turf. And the solution to our second riddle lies within this belt of limey rocks, near to Smoo Cave, by Durness.

Smoo Cave opens out to the sea close to where the Scottish coastline turns away from the Gulf Stream in the west and faces north towards the Arctic Circle. One blusterous day in 1989, my steel-framed tent was swept off the cliffs near here by a gust that whipped along the northern coastline, piling up the sand of the dunes. Mr Charles Peach had also been mapping in this gusty place during the summers of 1854 to 1857. After a successful bout of hammering, he sent his hopeful parcel of fossils from Smoo down to London for the attention of Mr J.W. Salter Esq.

When Salter looked at the fossils, he was in for a shock. In this package nestled a plethora of strange-looking snails and orthocone nautiloids. These newly discovered fossils bore all the hallmarks of others known only from Canada and the United States. In those distant regions, he knew that a similar succession occurs, which is broadly as follows: basal quartzite with 'pipes'—shale with 'fucoid' worm burrows—sandstone with *Salterella*—and then limestones with cabbage-like stromatolites, or shelly fossils, on top.[168]

Salter had stumbled on a clue to a big answer to a really big question, a full century before its final acceptance. The big question was this one: *are the oceans permanent features of the planet?* As shown later by the work of Alfred Wegener on continental drift and by Tuzo Wilson and others on plate tectonics—the answer was a loud 'No!' Ocean basins are mobile features of the planet, seldom lasting half a billion years or so. Salter's new fossil evidence, when assisted by later discoveries made by Ben Peach and others, would help to show that Scotland and Canada were not just sister nations. They had also shared a much longer prehistory, during which 'the Old Atlantic Ocean' had closed—to form the Moine Thrust—and then opened again—to form the modern Atlantic Ocean. Here, then, was evidence that an old ocean had disappeared.

A rock that swallowed time

To contemplate our third riddle, I drove the short distance along the road from Inchnadamph to Loch Assynt. Here can be found a splendid setting for our time travels, from the Cambrian explosion and far back towards some of the earliest signs of life on Earth. Standing along the shore of the loch, my eyes were drawn towards a neck of rock that juts out into the dark waters. This place bears the vivid scars of the Scottish Civil War—a tumbledown castle saddled with a mournful tale. Back in 1650, the loser of the Battle of Carbisdale, the Marquis of Montrose, had been imprisoned here, in Ardvreck Castle, to await his decapitation.

But it was not the neck of old Montrose that interested me. It was the fate of a fossil. One day, my son Alex had spotted a fossil sitting among beds of ginger-coloured sandstone near our feet, right by the shore of the loch. In fact he had spotted several fossils. Some were tiny and needed a lens to see them. Those looked something like the broken tip of a pencil and were called *Salterella* after J. W. Salter who first described them. Other fossils were somewhat larger—wiggles and bumps that resembled the windings of an ancient worm called *Planolites*.

Strangest of all, though, was an enormous 'trilobite' which we christened the Monster of Montrose. It was unusual because trilobites from the Cambrian of Scotland are seldom much bigger than a bumblebee. But this old rascal was huge—about the size of a Lochinver lobster, some 20 cm long. A closer look showed that it wasn't provided with a trilobite-like shell either. Instead it was preserved as a series of scratch marks, presumably left behind by the frantic digging of a trilobite-like creature in the mud, like those called *Rusophycus* from our Circus of Worms (see Chapter 4). This 'trilobite' had not only left us a ghostly impression of its spiny legs but also a forlorn message about its predicament. We were about 500 million years too late to save it.

The *Salterella* Grit on which we stood was easy enough to follow across the landscape. Weathering to something like gingerbread cake, this sandstone made a line of bogs and hollows across the moorland. Next, and just beneath it could be seen a bed of rock that bears the oldest trilobites in Scotland. Here, near to the shore of the loch, is where Peach and Horne had discovered *Olenellus*—the fossilized 'nephew' of *Fallotaspis*—in 1891.

Yet further west, and lower down the rock succession, lies the Pipe Rock with its Circus of Worms (see Figure 21). During his traverse in 1869, Charles Lyell was much delighted with them: 'Although this rock now consists of solid quartz, it is clear that in its original state it was formed of fine sand, perforated by numerous lob-worms or annelids, which left their burrows in the shape of tubular hollows . . . hundreds, nay thousands of which I saw as I ascended the mountain.'[169] But walking just a little further down the rock pile in the same direction, we find that the terrain changes markedly—these lower, older 'Torridon' sandstones are much coarser and redder than the ones above and the ground hereabouts seems less boggy.

Long ago, this hallowed patch of hillside around Carn Quinaig was traversed dozens of times over in search of clues to the deep history of life. The geologists, Messrs Macculloch, Sedgwick, Murchison, Nicol, and Lyell, all strode backwards and forwards across the heather here, notebooks in hand, no doubt frowning and scratching their heads. At first, it was thought that the bright white Pipe Rock passes down into the chocolate-red Torridon Sandstones without a big break. But that was to prove a serious oversight, as James Nicol was the first to discover. The break that he found between them—called an *angular unconformity*—was to carry with it a huge and somewhat terrifying implication. The red Torridon Sandstone had beds that were nearly horizontal. But the Pipe Rock on top was tilted at about 15 degrees to the east. This discordance was so angular that it pointed the way towards a passing of time of almost unimaginable largeness. Time had been swallowed up along a knife-thin boundary between the lower unit (the red Torridon Sandstone) and the one on top (the Pipe Rock). This angular unconformity therefore recorded a passage of time so great that the land surface had altered—by means of modest incremental change—in both its angle and its altitude (see Figure 21). Indeed, we now know that the Torridon sandstones are about 1000 million years old, while the Pipe Rock sandstones above them are about 530 million years old—a gap of nearly 500 million years.

Another huge gap had already been traced further to the west, by John Macculloch in the 1830s, following along the shore of Loch Assynt. Here, we could still see that the red Torridon Sandstone lies upon an uneven surface of crystalline 'basement rocks' that stretch far away westwards to the Isle of Lewis in the Outer Hebrides. These 'Lewisian' rocks are made of big old crystals with big old names—hypersthene and amphibole, orthoclase and plagioclase, epidote and muscovite. Even better, these rocks are now known to be among the oldest rocks anywhere on Earth.

This gap between the Lewisian crystalline basement—about 1000 to 3000 million years old—and the overlying red Torridon Sandstone—about 1000 million years old—and then the gap between that and the Cambrian Pipe Rock with its Circus of Worms—a gap of about 500 million years duration—was to help bring about a huge revolution in thinking. That is because these two gaps pointed directly towards the vast expanse of time that Charles Darwin needed for the modern theory of evolution by natural selection.

The Victorian stonemason Hugh Miller, writing in about 1841, was clearly caught by the romance of views around Quinaig:

> nothing can be more striking than the change which takes place in the landscape, in passing from the wild ruggedness of a [Lewisian] gneiss region to the level fields, swelling moors, and long undulating ridges of a [Torridon] sandstone one. But in the interior of the country, where the [Torridon] sandstone occurs

> chiefly in detached hills, it lends to the prospect features of surpassing boldness and grandeur. Rising over the basement of rugged gneiss hills, that present the appearance of a dark tumbling sea, we descry a line of stupendous pyramids from two to three thousand feet in height, which, though several miles distant in the background, dwarf, by their great size, the near eminences into the mere protruberances of an uneven plain. Their mural character has the effect of adding to their apparent magnitude. Almost devoid of vegetation, we see them barred by lines of the nearly horizontal strata. . . . and while some of their number, such as the peaked hill of Suilvein, rise at an angle at least as steep and nearly regular as that of an Egyptian pyramid, in height and bulk they surpass the highest Egyptian pyramid, in height many times. Their colour, too, lends to the illusion. Of a deep red hue, which in the light of the setting sun brightens into a glowing purple, they contrast as strongly with the cold gray tone of the gneiss tract beneath as a warm-coloured building contrasts with the earth-tinted street or roadway over which it rises. [The pebbles which it contains] are almost all water-rolled,—usually quartzose or feldspathic in their composition, though in considerable proportion jasperous . . . I failed to discover in them aught organic.[170]

Nowhere are these red Torridon Sandstones better seen than in the country around Loch Torridon itself.[171] Until 1856, these red Torridon Sandstones were still being compared with similar ones yielding fossil fish on the east coast, around Aberdeenshire and Caithness, which are a mere 370 million years old. But no fossil fish had yet come to light in them. Hugh Miller therefore spent a whole summer searching for them, but without success. And if Miller couldn't find his beloved fossils in these rocks, then probably nobody could. And nobody ever did. This reminds us of a peculiar dictum—a single fossil is worth more to science than a mountain of rock. It can change the whole narrative of Earth history.

The lack of fossils in the red Torridon Sandstones also presented Sir Roderick Murchison with a tricky decision. He was obliged to acknowledge that these old red beds could not be swallowed up into his expanding Silurian empire. Barren of fossils, Murchison tossed these Torridon red beds into his rubbish bin of stratigraphy—Sedgwick's Cambrian system. By 1888, the Cambrian was demarked by its own fossils—*Olenellus, Callavia* and so forth. Yet the old Torridon Sandstones still had none of these marker fossils. So Sir Archibald Geikie tidied up the stratigraphy yet again and tossed these tiresome red beds into the next rubbish bin of stratigraphy—the Precambrian. By that time, the Precambrian rubbish bin was starting to get a bit full. Nor was it truly without spectacular fossils in Scotland, as we shall soon see.

Darwin's Lost World—towards a first solution

Charles Lyell will have been acutely aware of this problem—the lack of fossils below the Pipe Rock—while he was climbing the pyramid of Quinaig in 1869. The crux of the problem both for him and for Darwin, remember, was this one. Although a few blobs and wiggles had turned up in rocks of broadly Cambrian age elsewhere, no geologist was pointing to any signs of life in the earliest rocks. Understandably, this riddle seemed to Darwin truly 'inexplicable', when he paraphrased his dilemma as follows: 'He [a critic] may ask where are the remains of those infinitely numerous organisms that must have existed long before the first bed of the Silurian system was deposited: I can answer this latter question only hypothetically . . .'[172] Some of his colleagues were minded to take all of this at face value. For Sedgwick and Murchison, for example, the appearance of worm tracks and trilobites in beds of the Pipe Rock could be taken to mark out Day One of life's creation. But like any good riddle, the statement of our puzzle—*that no geologist was pointing to any signs of life in the earliest rocks*—also contains its own solution.

The solution to this riddle lay not with *the signs of life*. It lay with the *pointing*. To cross over into this Lost World, investigators had yet to learn how to leap across a great divide—what we may call the 'visibility threshold'. This threshold divides a younger world to which we can point, because it is largely visible to the naked eye, from an older Lost World that is harder to spot because it was, and still is, microscopic. To leap across this divide, geologists had therefore to unleash that most devastating of weapons—the ground glass lens.

To look at the 'evolution' of the ground glass lens, we need to leave behind the rocks for a while, and to travel right down to the opposite end of Britain, indeed to Oxford, and take a gentle stroll down Broad Street. There, at the eastern end of this fine old thoroughfare, stand three imposing Palladian buildings which students cheerily ignore each day on their way to lectures. First of these is a kind of Roman temple, with steps and a portico. This is the Old Clarendon Building, where millions of Bibles have been printed over the years. It was here that Dr William Buckland also proudly displayed his bulging collections—fossils to the left, and minerals to the right. And it was here that Oxford students once took their oaths of allegiance, to protect the vast and ancient Bodleian Library from the process of oxidation—meaning, of course, the ravages of fire.

A little to its east lies that great ceremonial roundel called the Sheldonian Theatre. Resembling the famous Globe Theatre in London, but less draughty, the Sheldonian has been thoughtfully roofed over to keep Oxford dons from catching cold. Next door to the Sheldonian, and set back from the road behind iron railings, stands the Old Ashmolean Museum, now the Oxford Museum of the History of Science. Standing back from the street, it can appear somewhat lofty and aloof. A cascade of steps leads upwards to the great portal, beyond which lies a panelled room that still reeks of oak and beeswax. This room was once the setting for university meetings in all branches of Natural Philosophy, but now houses a collection of the earliest scientific instruments. Best of all, it contains some of the first ground glass lenses.

The lens is a typical product of the scientific world and hence of bottom-up thinking. And by that I mean a view of the world that sees *all* things as self-organized from the bottom upwards. In this world view, little things like protons and electrons and bacterial cells have given rise to big things like ocean gyres, geoids, and galaxies. But even big things, when they are far away, need the telescopic lens for us to see them. And little things, of course, need the microscope lens.

Oddly, such grand observations did not directly follow discovery of the lens itself. Magnifiers were being manufactured from quartz in its clearest form, known as rock crystal, far back in human history. Crystal lenses have been discovered, for example, in the graves of pyramid-builders in Egypt from about 2500 BC.[173] Roman writers also tell us that emerald crystals were used by the emperor Nero to spy upon his ballet dancers in about AD 65. It is not entirely clear, though, why these objects were not harnessed to a somewhat bigger question—*how does the world appear when we look at it in detail*? It may be, as the artist David Hockney has suggested, that lenses were kept hidden as secret knowledge—because 'knowledge is power'.[174] But it may also be that lenses were rare, or of inferior quality.

It was not until after AD 1500 that the ground glass lens came into the hands of a few brave and curious souls, such as Copernicus, Galileo, and de Leeuwenhoek. The invention of ground glass was seminal here because it led directly towards optics, microscopes, telescopes, and hence towards new ways of seeing. By 1650, treadle-powered lathes were available for more accurate lens grinding and microscope tube-making. And by 1665, Robert Hooke had used the compound microscope to begin his study of living and fossilized cells. Indeed, the word 'cell' was invented by Hooke. It was he who also realized that living and fossil materials were connected by a process we call fossilization. And Hooke was also the first to experiment with fossilization of shells and wood.[175]

I had come to the Old Ashmolean to look at microscopes because, among the old microscopes in the museum here, there is one like that used by Robert Hooke. It proved to be little more than a short stubby tube of cherrywood and leather attached to a brass stand. But with this strange device, Hooke and his peers had set in motion our major revolution in thinking. Before the invention of such glass technology, for example, 'bottom-up' scientific thinking was unavoidably restricted to armchair philosophers. But after the invention of the ground glass lens, and then of industrial lathes in about 1760, microscopes were no longer rare treasures, restricted to a few wealthy savants. As soon as glass technology, and industrial machining were in place, then lens-based 'bottom–up' thinking was able to bloom and to begin to transform our view of the world.

A slice of life

This transformation was further accelerated by the son of a Sheffield cutler with a passion for rock, called Henry Clifton Sorby. He revolutionized the way we think about rock, from the 1850s onward.

Using a battery of cutting wheels and grinding plates, Sorby showed that ordinary stone, when cut into slices like Parma Ham, can appear pleasingly vivid when viewed under the microscope with polarized light. From the 1870s, everyone wanted a polarizing microscope for their evening entertainment. For the rich, it was the colour television of its age. Certain rocks, and especially igneous rocks, were found to shine with a riot of psychedelic colour. Not only that, but the hues themselves, and the way they altered when the slide was rotated beneath the lens, were found to yield clues to both the minerals and the elements present.

It was therefore possible, from the 1870s onward, to decode the story of any rock from birth to burial. Alas, that was too late for Darwin and his first edition in 1859. But it was timely for the coming prophets of bottom-up thinking, like Thomas Henry Huxley.

HMS Globigerina

Not only was Huxley among the first to seize on the explanatory power of Darwin's big idea, he was eager to extol the virtues of microscopy to a public yearning for progress. Like Darwin, he had been a Ship's Scientist, on the voyage of *HMS Rattlesnake* in 1846. But whereas Darwin collected his best thoughts away from the ship—presumably to avoid sea-sickness—Huxley threw himself into the sea and marine biology.

Speculations about the nature of the deep sea floor were, in the 1850s, rather like those about 'life on Mars' today. They gave shape to questions about the habitability of our planet, and to the very nature of human existence. Dr Edward Forbes—discoverer of life in the Cambrian—had already begun to theorize about an absence of life in the deep ocean. For him,[176] modern marine life in deep water was very like that in the Silurian, with its biota of brachiopods and nautiloids, while life below the top few hundred metres of the modern ocean was lacking—azoic—just like the beginnings of the fossil record itself. If that were so, he pondered, could it be that soundings in deep waters would provide a mirror to help decode the deep history of life itself?

Darwin had drawn attention to the ideas of Edward Forbes, expecting that 'the bottom will be inhabited by extremely few animals' in 1859.[177] But by 1861, he had changed his wording significantly to say 'the bottom will be inhabited by few animals, but it will not be, as we at last know from the telegraphic soundings, barren of life.'[178]

The chance to test—and to reject—this big idea actually came in 1856, in the race to connect the Old World and the New by means of the electric telegraph. Huge cables of copper wire were manufactured in England and America, coated with insulation, and then laid by ships along the bottom of the Atlantic Ocean. But the engineers needed to know the depth and roughness of the seafloor. And that required deep soundings across the Atlantic—indeed, the first ones ever made. In one heroic journey of 1856, Lieutenant Berryman of the United States Navy gathered twelve samples from the deep seafloor and was amazed to find a great blanket of creamy grey sediment—now called *Globigerina* ooze (see Plate 13). From 1859, a flurry of ships began further exploration of the deep sea bed—each one sounding out the line, spinning the wire, laying the cable and then, of course, dealing with its breaking apart. Each time it broke apart, which happened not infrequently, the cable had to be hauled back on board, splashing *Globigerina* ooze across the deck.

Back on board ship, or in labs around the Atlantic, scientists like Ehrenberg and Bailey or Huxley and Carpenter, began to examine these 'deep sea soundings' under their compound microscopes, much as I later did aboard *HMS Fawn*. They found them to contain myriads of tiny creatures looking like raspberries, shuttlecocks, drain pipes and organ pipes, trumpets and flugel horns—all made by single-celled creatures such as diatoms and foraminifera. Soon it became apparent that this mini-menagerie mimicked that obtained by scrubbing a lump of chalk under water with a toothbrush. Indeed chalk, like that from the White Cliffs of Dover, was used as an early form of toothpaste. Musing on all this, Huxley delivered a popular public lecture in 1868—On a Piece of Chalk—and yet another in 1873—called Problems of the Deep Sea.[179] In these essays, he showed that the predictions of Dr Edward Forbes had been wrong. The deep sea was not barren at all. It teemed with life.

HMS Challenger

Much of what Huxley wrote was stimulated by another voyage of the Royal Navy—that of *HMS Challenger*. Once a ship-of-the-line, she was an eighteen-gun naval corvette of 2000 tons. In 1872, she had set out upon a three-year voyage, herself heavily loaded with scientists, and each scientist heavily loaded with books and beards. Their aim was to test this new idea—that the ocean seethed with life from top to bottom. The brainchild of Exeter Surgeon Dr William Carpenter, this splendidly hirsute expedition gathered an inventory of the Earth's biosphere that was to be like no other before or since. It was to be the Apollo Mission and the moon rock of its age. For the century that followed, be-whiskered microscopists were to cut their teeth on microbes wrested from the deep by *HMS Challenger* and her successors.

One day on *HMS Fawn*, we therefore decided to put the findings of the *Challenger* mission to the test. We maintained a stationary position some two miles off the coast of Barbuda, on its eastern side. That is where the volcanic basement of the island slopes rapidly down towards the deep Atlantic floor, like the setting for an undersea toboggan race. At this short distance from shore, the white line of surf that marked the reef could still be seen dancing on the skyline. But the water beneath the ship was here already 1600 metres deep.

For the previous eight months, our donkey engine had been sitting idly on the fo'c'sle deck. It had been quietly annoying Captain Lou Davidson because of its oily exudations, which soiled his teakwood decks, which then spoiled the jib of his ship. Now, however, its little moment of glory came and we primed up the diesel engine. On went the core tube, belayed at the end of a wire. Not any old wire, of course, this cable was over a mile long. The engine then sputtered into life and we paid out the wire, for hour after hour. At long last, the time came to haul it back in and see what we had caught. Tension mounted as the core itself slowly scaled the ocean stairway, to return to its rightful place on board ship.

When a sampling device breaks through the surface of the ocean from the deep seafloor, there is a strong mixture of emotions among the crew. The first is joy—like that of seeing a dear friend returning from a long and dangerous adventure. The second is anxiety—that the device might come back empty handed. But the most potent is fear—that the device itself will do mischief, or even great injury, to those on board. This happened earlier in our mission on *HMS Fawn*, while crossing the Atlantic near the Azores. We were not only 'underway'[180] but we were sampling the deep seafloor as we went—with a device called an underweigh sampler. This bounced a scoop at the end of a cable. Unhappily, during the darkness of the night, the little red flag that announced its imminent arrival on board ship passed through the gloom unseen, and the sampler hit the davit—a big iron fishing rod—at such speed that it snapped the cable. The cable of taut wire then lashed across the deck, taking the scalp off a poor Able Seaman, with terrible injury. Mercifully, nothing so vexing was to disturb our sampling of the island slope. The creamy grey ooze was quickly gathered up and examined under our waiting microscopes. We could now see what Dr William Carpenter had marvelled at—microbes—millions and millions of them.

One of these microbes was that little protozoan that still sails the surface of the sea, called *Globigerina*. Shaped like a miniature raspberry, less than half a millimetre across, it sports ten or so tiny chambers coiled around an axis like the slide of a helter skelter. There are three globular chambers for every whorl around the spire, the whole 'ship' being inhabited by a single celled protozoan, much as with the other foraminifera we have met around Barbuda. *Globigerina* is now known to sail across the upper layers of the ocean in its trillions, catching small shrimps with sticky fingers called pseudopods. These little arms can stretch out into the brine like the lines of a fishing rod.

At first, it was thought that *Globigerina* and its relatives lived on the deep seafloor. Thomas Huxley famously championed that view in his lectures, that is, until Dr Wyville Thompson—one of the Great Beards of the *Challenger* expedition—showed that *Globigerina* was also turning up alive and well in plankton nets trawled across the surface of the ocean. This little foram, it seems, had merely fallen from the surface on to the deep seafloor once its tiny shell had been discarded like a used condom. That analogy is chosen with care. Since every particle of a foram's soft body takes part in sex, its unwanted shell is emptied and can fall to the seafloor. Henry Brady and others were quick to observe that *Globigerina* was a bit unusual in this respect. The vast majority of foraminifera brought back to the surface by *HMS Challenger* had actually been living directly on the deep seafloor—until, that is, they were so rudely interrupted.

With foraminifera abounding in the deep—supposedly the abode of the most primitive life—it was therefore only a small step towards looking for foraminifera in the earliest rocks themselves. Unhappily, this led William Carpenter, and then Charles Darwin, towards a glutinous trap—like the sticky fingers of a giant foram—called 'The *Eozoon* Debate.'

The *Bathybius* mystery

To understand the *Eozoon* debate we need first to inhabit the thought-worlds of both Huxley and Carpenter. Both were committed evolutionists, of course. But neither seem to have been committed 'Darwinists' in the way we would now understand it. That is to say, neither were convinced that Darwin's big idea, that '*evolution by natural selection is the thing that explains all of life and biology*', was a wholly satisfying explanation for the history of life. Indeed, through most of the later nineteenth century, it seems that true 'Darwinism' was quite out of fashion.

But that is not to say that evolution itself was not coming firmly into fashion. Evolution took the world by storm, shaking the Establishment, and giving new shape to the dreams of working people and society at large. The question here was really this one: did evolution work in a democratic and unpredictable kind of way—as Charles Darwin had suspected? Or was evolution rather predictable, following a set of already written blueprints—programmed evolution as we would now call it. That was the view championed by German biologist Ernst Haeckel and it also became the view of Dr William Carpenter.

During the 1870s, there was also a feeling 'in the air' that deeper and deeper soundings from the sea would bring to the surface ever more primitive creatures, like sponges and foraminifera. This train of thought is marvellously illustrated by the curious case of *Bathybius*. By 1868, some of the original deep sea soundings had been sitting around in jars on their shelves for a decade. When Thomas Huxley came to inspect one of his old samples, he was fascinated to see that the jar contained a slime—like the white of an egg but criss-crossed with veins. This slime had not been there when he first examined it in 1857. Thinking that he had discovered evidence for the most primitive kind of life—without cellular structure—he christened it *Bathybius haeckeli*. This name was given partly in reference to its deep sea habitat—the *Bathybius* bit, and partly in honour of a German Professor—the *haeckeli* bit. Happily, Ernst Haeckel was not the slightest bit offended by being compared with a jar of deep sea slime. Far from it. *Bathybius* was exactly what he had been expecting to turn up in the deep sea, on the boundaries between non-life and life. Happily, too, German is a peerless language for the describing of slime. He called it *Urschleim*, the Mother of All Slime.

Poor Huxley was in for a bumpy ride, though. The Great Beards of the *Challenger* expedition were hugely unimpressed with his *Bathybius*, thinking it a kind of fungal infection, perhaps like athlete's foot, or—horror of horrors—a product of chemical reactions within the jar itself. Unhappily, that is exactly what it turned out to be—mineral slime brought about by reactions between seawater and its preservatives. We can still see something like this forming in an old jar of pickled beetroot today. Poor old Huxley was clearly in a pickle, too. So he then showed us what a great scientist will always do when faced with a great mistake—he publicly apologized and recanted his error, in 1879. Ernst Haeckel, however, had less room for manoeuvre. He didn't withdraw from the bathos of *Bathybius* until 1883.

During all of this debate, there seems to have been an expectation that new monsters from the deep would confirm what we may call a 'programmatic hymn sheet' for the evolution of life. The first reports from the *Challenger* seemed to confirm this. But there was actually another strange expectation too—that the most primitive 'animals' alive today—like sponges and foraminifera—had barely evolved at all during the vast expanses of Earth history. William Carpenter, for example, who was an expert in these matters, seemingly saw in his forams some kind of 'Peter Pan of the deep sea world'.

This view became unsustainable, of course, once the dossiers on the *Challenger* expedition started to rattle off the steam-powered printing presses in London. Indeed, it was later learned that *Globigerina* and its relatives were not really primitive at all. They had evolved almost as fast as mammals. They can even have more chromosomes in their genome than we humans have in ours. But in 1859, none of that was known. And the mighty Mofaotyof was about to strike.

Eozoology

At about the moment when Charles Darwin was burnishing the last sentence for his book in 1859, Sir William Logan in Canada was polishing off 'a first' for the book of life—his candidate for the world's oldest fossil. Called *Eozoon Canadense*—the dawn animal of Canada—it resembled a Greek baklava cake in which the 'flaky puff pastry' was replaced by green serpentine, and in which all the 'honey-filled spaces' were filled with calcite (see Plate 13). This visually arresting, not to say mouth-watering, structure had been found only the year before by a Mr J. McMullen on the banks of the River Ottawa to the west of Montreal. Sir Charles Lyell can provide us with a description of it as it seemed to him in 1865:

> It appears to have grown one layer over another, and to have formed reefs of limestone as do the living coral-building polyp animals. Parts of the original skeleton, consisting of carbonate of lime, are still preserved; while certain interspaces in the calcareous fossil have been filled up with serpentine and white augite. On this oldest known organic remains Dr Dawson has conferred the name of Eozoon Canadense; its antiquity is such that the distance of time which separated it from the Upper Cambrian period . . . may, says Sir W. Logan, be equal to the time which elapsed between the Potsdam sandstone [Upper Cambrian, now dated at about 490 million years old] and the nummulitic limestones of the tertiary period [such as the pyramids and sphinx of Egypt are made from, which are dated at about 50 million years old][181]

Eozoon was discovered in the middle of Canada, in a bed of marble some 500 metres thick that was intermixed with thick layers of banded gneiss and micaceous schist. Now marble is known to be a highly squeezed and altered form of limestone. That squeezing is what gives it the attractive swirls and mottles. In 1864, this strange kind of marble was shown to Dr J.W. Dawson of Montreal—a one-time pupil of Charles Lyell—who was so taken with it that he named it *Eozoon Canadense*. Dawson was equally intrigued by the abundance of carbon, in the form of graphite, to be found in the rocks containing *Eozoon*. Not unreasonably for his time, he inferred that this pointed to some kind of vegetation long ago. But what kind of vegetation he could not say:

> There is thus no absolute impossibility that distinct organic tissues may be found in the Laurentian graphite, if formed from land-plants, more especially if any plants existed at that time having true woody or vascular tissues; but it cannot with certainty be affirmed that such tissues have been found . . . and when we think of the great accumulations of Laurentian carbon, and that we are entirely ignorant of the forms and structures of the vegetation which produced it, we can scarcely suppress a feeling of disappointment . . . It may be that no geologist or botanist will ever be able to realise these dreams of the past. But on the other hand, it is quite possible that some fortunate chance may have somewhere preserved specimens of Laurentian plants showing their structure.[182]

So the material with *Eozoon* seemed much more promising. It was duly taken on tour by Sir William Logan, to be displayed before a curious audience at the Geological Society in London. There, the eminent microscopist William B. Carpenter was struck by a seeming resemblance between ancient *Eozoon* and some living foraminifera, like *Homotrema*. As we have seen in Barbuda, 'forams' such as *Homotrema* can be massively abundant and rock-forming. Indeed, one of these rock-forming 'forams' was even called *Carpentaria* after Carpenter himself. The white layers of marble were therefore regarded, by both Dawson and Carpenter, as the remains of skeletons that had grown, layer by layer, to contribute towards great reefs of limestone. These layers were then thought to have been infilled by a waxy green mineral of a mottled, snakeskin appearance, appropriately called serpentine.[183] There was a dreadful misunderstanding here, though. *Eozoon* wasn't a 'foram'. Indeed, it wasn't even a fossil.

But *Eozoon* did, at least, come from deep down in the azoic—in rocks then called Laurentian—the crystalline basement of Canada. And north-west Scotland. Today, we know that the rocks with

Eozoon are of a highly respectable age—about 1100 million years old or more. In other words, they were broadly coeval with the Lewisian rocks we have just seen near to Quinaig. Indeed, examples of *Eozoon* were later to be found—and sold to Victorian collectors for a handsome profit—from marble quarries at Led Beg nearby. But that should have given a clue too. These marbles are of Cambrian age. And they were produced near the plane of the Moine Thrust. The *Eozoon* rock was the outcome of a mountain that stood on its head.

Hugh Miller, in his travels, had come across these marbles around Inchnadamph, long before *Eozoon* was found there. But my own first sight of such fancy marble came to light during field work at Dornie, a little further south. We had been set the task of mapping the Lewisian which, as we have seen, is some of the oldest and deepest crystalline rock in the world. When we reached these rocks, in September 1967, the hilltops thereabouts recalled a lunar landscape—a mix of bumps and holes. The bumps were the metamorphosed Lewisian complex, schist and gneiss. And the holes were small lochs, called locharns. Happily for mapping, each bump and locharn had its own strange shape. But not only did this place look like the surface of the moon, Tony Barber was training us to map it as if it *was* the surface of the moon—using aerial photographs—but with a subtle difference, of course. We could at least touch the rocks.[184]

With maps in hand, we would race up the scarp each morning, to reach the eerie plateau of gneiss and schist. Each day would find us hiking from bump to bump through bog and mist. Little by little, we traced the outlines of those rocky outcrops on to transparent overlays, using a compass to plot our position. Piece by piece, the blobs of pink and blue, green and yellow started to meld together into something we rather hopefully called a geological map. Only then did we see a strange phenomenon emerge—a wisp of greenish marble winding across the landscape, set within a field of glinting gneiss and twinkling schist. This was like a layer of *Eozoon* marble. And it had been caught up within the ancient gneiss like a swirl of butter within a bowl of Scots porridge oats.

We could see that these ancient rocks with the *Eozoon* texture had been squeezed and cooked to the point where nothing remained of their homely ancestral features. This swirl of marble had once been a continuous sheet that was later pulled apart by massive, mountain-building forces. No ripples, no pebble beds, no limestone and no lava remained. This was no laughing matter, either, for those in the search for the earliest signs of life.

Decoding of such 'metamorphic' rocks was to remain enigmatic—largely guesswork—until about 1880. It was only then that the real advances came, following hard on the heels of the petrographic microscope, the polarizer, the analyser, the rotating stage and the diamond wheel for rock cutting. Thus it was, for nigh on twenty years after publication of the *Origin of Species*, that decoding the oldest crystalline rocks—and hence of *Eozoon*—was a very chancy business. And Principal Dawson of Montreal took his chance.

With Dawson adamant about *Eozoon* and Carpenter enthralled by its resemblance to a foram, both Lyell and Darwin were caught in a web they could not see. Darwin was therefore moved to write, in the later editions of the *Origin of Species*:

> and the existence of the Eozoon in the Laurentian formation of Canada is generally admitted. There are three great series of strata beneath the Silurian system in Canada, in the lowest of which Eozoon is found. Sir. W. Logan states that their 'united thickness may possibly far surpass that of all the

succeeding rocks from the base of the Palaeozoic series to the present time. We are thus carried back to a period so remote, that the appearance of the so-called Primordial fauna (of Barrande) may by some be considered as a comparatively modern event.' The Eozoon belongs to the most lowly organized of all classes of animals, but is highly organized for its class; it existed in countless numbers and, as Dr Dawson has remarked, certainly preyed on other minute organic beings, which must have lived in great numbers. Thus the words which I wrote in 1859 about the existence of living beings long before the Cambrian period, and which are almost the same with those since used by Sir W. Logan, have proved true.[185]

The stage for our tragedy was now set. The main problem was the inferred but mistaken similarity between *Eozoon* and complex living foraminifera such as *Homotrema*. This led both Carpenter and Dawson, to conclude that foraminifera had barely evolved since the Laurentian period.[186] But then a fresh set of observations helped to sort matters out—*Eozoon* was not a *fossil* at all. First to strike were two Irish geologists—William Kind and Thomas H. Rowney—in 1866. Next to denounce the fossil was the German microscopist Karl Möbius, in 1879. A firing squad of specialists soon lined up—mustering their shiny brass microscopes—to condemn *Eozoon* as little more than a mineral growth, formed at great depth and high temperature. In the Assynt region, it was found next to major faults and intrusions. And in Italy it was seen coming out of a vent in Vesuvius! Soon enough, therefore, the game was up. All agreed that William Carpenter was a fine microscopist. Indeed, he was clearly something of a genius—a founder of modern psychology, comparative neurology, and of microfossil microscopy. But he was, alas, too far ahead of the game in metamorphic petrology. And he was therefore doomed to fall into the dreaded Mofaotyof trap, like many others before and since.

Carpenter never repented. In 1885 he was writing up his thoughts on the matter of *Eozoon* when another tragedy struck. Given to bathing in the evening by the light of a spirit lamp, he accidentally overturned it, setting himself ablaze in the bath. Like his old companion *Eozoon*, poor William was fatally transformed by heat. It was a cruel conclusion to a magnificent career.

Bottom-up thinking

How did life begin? Nothing in the realms of science has yielded such a 'big question' as this. And science, which is a unique system for the measurement of doubt, always rejoices in a good question. Our minds are always seeking explanations for the inexplicable. This 'Mother of All Questions' also spawns a series of equally interesting daughters. *Was the emergence of life on Earth easy or difficult? Are we alone in the Universe? And why are we here at all?* Questions like these are especially hard to answer. That is how they manage to stay both old and fit, spawning many conflicts along the way.

The answer to this howdunnit—*how did life begin?*—really matters to us now because it helps to define the nature of the human condition. Even in science, however, big questions like these can appear to have more than a single answer. This is awkward because the answers to big questions affect us deeply. They have great predictive power. We are all trying to guess what lies over the hill, for us and for our children. If we guess the wrong answers, we could well affect the fortunes of civilization. When Don Cortes and his men arrived in Mexico, for example, the Aztec soldiers greeted them as gods—but would they not have done better to fight them as enemies? Or when the great Christmas

Tsunami struck Asia in 2004, should the tourists have run inland or simply stood on the shore and prayed? And now that AIDS is striking in Africa, should doctors inoculate against the virus or invoke all the Angels in Heaven? There is no doubt about it. When it comes to the crunch, *seeing the world as it really is* will matter to us, and to our children, very much indeed. And that means gaining a proper understanding of the *Bathybius* and *Eozoon* debates.

The lessons to be learnt from these episodes are worth a little reflection before we set out upon the road again. Most striking was the prediction that early life would be like those things found living in the deep sea today. Such a mindset or paradigm still exists in various forms all around us. Paradigms are the maps of science. They give shape and substance to our questions. But, as we have seen, these maps can be fiercely promoted and just as fiercely abandoned if they fail to predict what lies over the hill ahead. Hence, our tests have tended to get better and stronger with time—from the ground glass lens to the electron microscope, and from armchair reasoning to the ground-truthing of a geological map.

We can accept that the unfortunate *Eozoon* was a classic example of Mofaotyof thinking—make your fossils seem as old and as interesting as you can. But we must also acknowledge that both Dawson and Carpenter were not fools. They were among the best and most innovative microscopists of their age. Together, they put forward a very strong idea—that life began with a kind of giant protozoan. Their idea was so clear and so elegant that it was simple enough for colleagues to test it. And that is exactly what their colleagues did. To destruction. Nor was *Eozoon* really bad science. 'Bad science' consists of ideas so vaguely formed that they can only be weakly tested, or of ideas that cannot be tested at all.[187] But the ideas of William B. Carpenter were arguably examples of good science at work and play—strong and playful ideas that could be—and were—strongly tested.

By 1907, *Eozoon* had largely disappeared from the liturgy of biology and from the popular literature too. But its disappearance did not require a return to that earlier statement—'*that no geologist was pointing to any signs of life in the earliest rocks*'. Stranger things were about to emerge.

CHAPTER NINE

TORRIDON

Water of Life

At long last, I was about to approach the pinnacle—a lonely pyramid of rock called An Teallach. This is just one of a series of peaks that stretches from Loch Assynt down to Loch Torridon, in the north-west Highlands of Scotland. Each mountain is like a monument, standing above its own plain of bedrock, provided with its own distinct personality, isolated and aloof. Had Henry Moore ever sculpted mountains instead of mother-goddesses, he would have made them look like these—Quanaig, Suilven, Canisp, and An Teallach. Crofters hereabouts once thought of them as the haunt of fairies and giants, and it is easy to see why. This is a place without equal—surely the finest spot on the planet for contemplating the realm of Darwin's Lost World.

As the sun began to sink, the pinnacle of An Teallach became cloaked in a veil of mist. Next morning however, the veil dispersed and my little research group decided to clamber up to the peak. The path climbed ever more steeply through layer upon layer of flinty pebble beds, allowing us to share with Darwin his great awe at such a sight: 'In the Cordillera I estimated one pile of conglomerate at ten thousand feet in thickness.'[188] These red beds—the Torridonian sandstones—are easily as thick, and they decorate the cliffs by the path up the mountain with stripes as clear and sharp as any on a tiger's back. We knew that each horizontal stripe concealed the history of an ancient river bed that had woven backwards and forwards across this landscape, not just once but thousands of times. Indeed, so often had the rivers swung across this land that river pebbles were stacked neatly in layers to build a package nearly four miles thick. When I sat down to catch my breath on the way up, I tried to imagine what it might have felt like to sit beside one of these ancient rivers—a harsher world of sand and rock, punctuated by thunderstorms and flash floods. I tried, as well, to contemplate the vastness of geological time so boldly shown by the cliffs. Accustomed as I was to such a game, my brain still boggled at the enormity of it all.

Climbing near to the top, we could begin to see, to the north-west, the crazily indented coastline of western Scotland with its sprinkling of tiny islands. Far to the north stood the dim outline of Quinaig and those rocks that swallowed time. Closer to hand, we could spot the glistening Moines and the mountain that stood on its head. The barren Moines seemed to be leaning heavily on fossiliferous Cambrian below, as though trying to teach those young upstarts a lesson. And over there, beside Ullapool, could be seen the Durness limestone, left behind by an ocean that had disappeared.

The rocks immediately before us were very old indeed—they had sat uncomplaining through as little as one billion to as much as three billion summers. The most ancient of these were crystalline gneisses—hoary old Lewisian rocks. These were almost as elderly as those from the Pilbara region of Western Australia.[189] But the Lewisian rocks of Scotland have been so cooked and squeezed that they now bristle with silvery mica and wine-red garnet, set within a mass of grey quartz and ivory white feldspar. The scabbard of Excalibur, King Arthur's trusty sword, with its iron inset with silver and garnet, could hardly have shown anything more noble.

An even finer treasure awaited us, though, as we ambled slowly along the shores of nearby Loch Torridon. Down by the water's edge, we played at being time travellers, wandering through an ancient landscape of lagoons. Those dark and bulky slopes of Lewisian gneiss above us had been cut by wind and rain long before the evolution of jellyfish and worms. And the little valleys, now filled with pebble beds, had once been the home of rivulets that babbled over a billion years ago. Even modern Loch Torridon now chose to slumber on the floor of a much older lagoon, which I shall here call Old Lake Beatha—the ancient water of life. As we ambled along the edge of the modern loch, we could see that these billion-year-old lagoonal shores had been washed by waves, baked by the sun and then wetted by raindrops, much as with Loch Torridon today. We had arrived too late to feel the mud squelching between our toes. And too late to smell the methane bubbling up from below. But the question in our minds was this—had we arrived too late to resurrect the lives of Beatha's long lost creatures?

Here, then, was a landscape to compare with any from ancient myth. And like King Arthur of old, waiting on his distant island of Avalon, this landscape seemed to be brooding as though lying in wait for something to happen.

The Crone Stone

Perhaps the finest gem in this landscape was set further offshore. We knew that this little gem was likely to be found within a bluff of stone—rather like Arthur's sword. And that it had something to do with a fabled lake—also rather like Excalibur. But we did at least have some ideas as to where it lay—exposed on a remote crag known as the Old Crone's Head.

To reach this crag, we would need to make a journey by boat. A passage to the outcrop was therefore negotiated with Bill, a local crofter. For a spell, we waited on a remote peninsula overlooking the sea loch. Then we heard the buzz of an outboard rising above the splash of the waves and saw an orange boat bobbing towards us. After loading up with post and supplies, we jumped aboard and pushed away from the shore, to race over the choppy waters towards those distant cliffs. Because of wind and tide, we arrived on the opposite shore damp, bedraggled, and dazed, and some way short of our destination. After offloading the post and the camping gear, we therefore set off on a hike across some miles of rugged moorland.

A blustery wind accompanied our hike but it progressively turned to drizzle and then to a clammy coastal fog. Through it, we peered hard into rock pools, scraping aside the drapes of barnacles and bladderwrack, limpets and slime. I knew well enough what to look for here—it would seem like little more than a dark smudge. But I had never looked for it in rocks so old. There at last it was, though—a pebble of phosphate concealed within layers of ancient lake bed. Out in the wind and the rain, it looked small and insignificant. But that was part of its secret. We carefully wrapped it up and took it, along with some of its fellow pebbles, back to the Oxford Palaeobiology Labs. Our aim was to coax these rocks to give us the missing clues to the game of Darwin's Lost World.

Gaelic amber

Back in Oxford, we set about cutting up slices of the phosphate to examine them under the microscope. Within a few days, with help from Owen Green, we were able to scan them for fossils. The best of these slices felt like amber, rather waxy to the touch. Indeed, they even looked like amber—a golden honey colour when held up to the light. But it is important to remember this. The oldest amber with inclusions is a mere 140 million years old, from the age of the dinosaurs. Our little phosphate beauties were almost ten times as old. And they were filled with tiny treasures ten times as strange.

Anxiously, I zoomed in with the high power lens of the microscope to see if anything was still 'at home'. This is always an exciting and rather tense moment. And, yes indeed, there was something still at home. Down the barrel of the viewing tube—and displayed on the computer screen nearby—could be seen a cluster of cells so finely preserved that their frilly sculpture and yolk-like contents stood out bold and clear (see Plate 14). But that was not all. This little cluster had been 'caught in the act' of going about its daily business. It had been escaping from a safe hiding place. Looking rather like a plastic bag torn apart in a storm, this hiding place of waxy material—called a resting cyst—had probably provided a home-grown haven through the winter months a very long time ago. How long or how dark those winter months were, it was hard to say. But it seemed reasonable to conjecture that this little cell cluster had begun to burst out of its bag during the warmth of spring—only to fall into the honey-trap of fossilization.

Eager to see how these microfossils appeared in three dimensions, I clicked the computer mouse to bring up some of the fancy software. In no time at all, this tiny menagerie seemingly leapt into life. Scanning across the slide, I could also see how this accident of fossilization had been repeated many times over on the bottom of old Lake Beatha. The golden brown slice of rock was as packed with life as an Oxford pub on a Saturday night. Some looked like groceries—cells bursting out of bags like apples, called *Torridonophycus*; others were linked up like sausages, or stacked up like biscuits, and some were collected together in blocks like a sachet of Aspirin tablets (see Plates 15 and 16). There were things here like bunches of grapes, wine bottles, and tiny silk stockings, all abandoned on the ancient lake floor as though after a riotous weekend.

Dream come true

We were not, however, the first to contemplate the potential of ancient phosphate for the detection of early life. Far from it. Indeed, my hair almost stood on end when I came across the following words concealed within the dense verbal foliage of Darwin's *Origin of Species*: 'The presence of phosphatic nodules and bituminous matter in some of the lowest azoic rocks probably indicates the former existence of life at these periods.'[190] Charles Darwin was here implying that phosphorus and carbon are the essential building blocks for life, both today and in the distant past. Concentrations of these elements in very old rocks are therefore consistent with life extending far back beyond the Cambrian explosion. Here, then, was a potentially illuminating idea. Darwin had strangely provided us with a clue that could help to solve his very own puzzle.

Unfortunately, Charles Darwin left us no clues about the source of his 'phosphatic nodules'. Our nodules from Scotland could have been known to him but we cannot be sure of that. Even so, these Torridonian nodules can lay claim to another major scientific distinction. They were the first rocks of any kind to produce high quality cellular fossils from before the Cambrian period. Geological surveyors Mr Ben Peach and Mr John Horne had spotted the nodules in the crags near the Old Crone's Head by 1899. Even more importantly, both Peach and Horne concurred with Darwin about the significance of phosphate as a signpost for early life, and then they had the wit to send them off for microscopic analysis by J. J. W. Teall, who first reported on these remarkable fossils as early as 1902.[191] It was really Jephro Teall's painstaking attention to detail that made the breakthrough, of course. His pioneering work with the microscope provided something like the deep orchestral colour that was to accompany that great geological symphony we call the *Memoir on the Geological Structure of the Northwest Highlands of Scotland*, published in 1907.[192]

Curious to report, our Scottish pioneers—Peach, Horne, and Teall—can now be seen to have ventured a little too far ahead of the field. Both fossil cells and fossil cells in phosphate were dangerously new ideas. Nobody—including themselves—really knew quite what to make of it all. With the ghost of *Eozoon* still lurking in people's minds, most scientists decided to let things lie fallow for a while. In no time at all, it was 1914 and phosphorus was quickly to turn from being an object of contemplation to being a weapon of war.[193] Almost fifty years were therefore to pass by, before remarkable preservation in the Precambrian began to make the news again.– this time from the Gunflint Chert of Canada and discovered by Elso Barghoorn and Stanley Tyler.[194] And by that time, both Peach and Horne and Jephro Teall were long dead.[195] The British Empire was on its last legs too and America was in the ascendant. Almost everyone therefore forgot about the Torridonian and its ancient fossiliferous phosphate nodules.

From cells to dung

As we have seen, this clue to the missing history of life—Darwin's Lost World—lay concealed within a phosphatic nodule. But when

Darwin wrote about phosphate and the search for the earliest life, could he have ever anticipated that phosphatic nodules could preserve cells so well? That may sound unlikely to us now. Darwin had, as we have already seen, written that 'No organism wholly soft can be preserved'.[196] But there is also reason to think he would not have been so very surprised, either. One clue to this occurs in his 'Journal' from the *Beagle* voyage of 1831–7, concerning the remarkable preservation that he saw in much younger cherts from Chile: 'how surprising it is that every atom of the woody matter . . . should have been removed and replaced by silex so perfectly, that each vessel and pore is preserved!'[197]

As a Victorian gentleman who owned both a pigeon loft and a vegetable patch, Darwin will also have known about animal dung. It is, of course, very rich in phosphate. In fact, animal tissues often tend to be richer in phosphate than those of plants. That may be because animals move about more, and so need more ATP (adenosine triphosphate) to act as a storage-battery within their cells. Darwin also certainly contemplated the role of dung in the formation of phosphate, and even the potential of phosphate to preserve organic matter. Here, for example, is some of his writing on bird dung phosphate—called guano—encountered by the crew of *HMS Beagle* on a lonely crag in the middle of the Atlantic:

> The rocks of St Paul appear from a distance of a brilliantly white colour. This is partly owing to a coating of a hard glossy substance with a pearly lustre, which is intimately united to the surface of the rocks. This [guano], when examined with a lens, is found to consist of numerous exceedingly thin layers, its total thickness being about the tenth of an inch. It contains much animal matter, and its origin, no doubt, is due to the action of rain or spray on the birds' dung. . . . When we remember that lime, either as phosphate or carbonate, enters into the

> composition of hard parts, such as bones and shells, of all living animals, it is an interesting physiological fact to find substances harder than the enamel of teeth, and coloured surfaces as well polished as those of fresh shells reformed through inorganic means from dead organic matter—mocking, also, in shape some of the lower vegetable productions.[198]

But while Darwin's own example of guano is certainly interesting, it arguably doesn't help our story in deep time very much. That is because phosphate in the form of bird guano doesn't go back through the fossil record for much more than 50 million years. Nor does it show, when we look at it under the microscope, anything more than the mashed up remains of organisms—shells, bones, and seeds—plus the ghostly outlines of bacteria. There were, and still are alas, few beautifully preserved cells in bird guano.

Travelling a little further back from the voyage of the *Beagle*, we meet again with William Buckland of Oxford. In the early 1800s he was turning fossil dung into yet another of his many obsessions. It is important to remember that Buckland lived at about the same time as the poet William Wordsworth, the painter John Constable, and the pianist Franz Schubert. It is no disgrace to observe then, that Buckland was a Romantic geologist in the way that Wordsworth was a Romantic poet, sharing a deep love for the landscape. By all accounts, though, Buckland was also a bit like an overgrown schoolboy, delighting in anything disgusting—serving up 'bluebottles' for dinner, feeding a pet hyena under the table, swallowing a royal heart for a joke; and filling up his rooms with fossilized dung from Jurassic reptiles. No surprise, then, that students are said to have flocked to his lectures like flies. Clearly, he was jolly good fun to have around. But for William Buckland, the pleasure as well as the problem with coprolites—the polite name for fossil dung—was one of image. True, they do improve somewhat with the passage of time. But they don't look particularly charming, even when mounted on a stand. It was only later, when scientists started to cut them up into slices, that they started to show some of their beautiful inner secrets. Once they were sliced up, as by the followers of Nicol and Sorby, such microscopists could begin to see little bits of shell, sponge spicules, and clots of bacterial matter. In other words, examination of Dr Buckland's coprolites—or perhaps we should say, those of his pet hyena Billy—would reveal, even after a few years, the mangled remains of something like a full English breakfast: eggshell and hogs bristle, tea twigs and oat husks. But we need to note this, too. There were, and still are, few beautifully preserved cells within either dinosaur doings or hyena dung. The bellies of beasts have truly 'done them in'.

This picture changes, however, once we decide to travel backwards in time from the Jurassic to the Cambrian. A transformation takes place from the modern world of *dung* towards an ancient world of *cells*. As we have already seen in the Cambrian rocks near to Emei Shan in China, we can find the so-called fossilized jelly babies and even the remains of invertebrate animal dung. But it is in those rocks from before the start of the Cambrian that phosphate starts to get really interesting. Fine examples of this were first shown to me by my Chinese student Zhou Chuanming and later by my American student Maia Schweizer. Both had been collecting phosphate rocks from the *c.* 580 million-year-old Doushantuo beds in the Hubei and Guizhou provinces of south China, with the kind help of Shuhai Xiao, Chen Junyuan, and Sun Weiguo, and brought them back to Oxford for study.

The best place for collecting phosphate fossils, it seems, occurs in the great phosphate mines of Weng'an, in Guizhou province to the north of the great Yangtze river. But access to these great mines is not easy. Visits still require the filling in of a mound of paperwork, penetration of several tough security barriers, and being accompanied by a police escort. There are at least two explanations for all of this. Either these mines are a penal colony for highly dangerous convicts. Or the fossils are being unusually well looked after. Hopefully, the latter.

To examine the Doushantuo rocks back in Oxford, we dissolved the rocks in dilute acid and then perused them under the scanning electron microscope. What we saw was very different from our Cambrian fossils from Emei Shan. There was nothing that looked like teeth, bones, shells, or dung. Instead, the phosphate took the form of beautifully preserved clusters and sheets of cells that we might dare to call fossilized seaweeds. Some of these cells even resembled tiny soccer balls—spherical clusters of cells with tightly-fitting sutures like the patches on an old leather football. Some of the footballs were made up from just one or two cells; others from between four and thirty-two tightly packed cells. It seems likely these footballs were the growth stages of various cellular organisms that had been programmed to subdivide themselves progressively into ever smaller cells—by something like the so-called 'reductive divisions' seen in an embryo.[199]

Having encountered fossils like footballs, it should not come as a surprise that they have been kicked around a bit in a great game of palaeontological soccer. Defending the goal mouth has been the role of the home team—of Shuhai Xiao, Chen Junyuan, and others. Their team argues as follows: since these footballs can look much like the stages in the development of an animal embryo—from an egg to a blastula to a gastrula—they provide us with unequivocal evidence for the earliest known animal cells in the fossil record. But, as we shall see, it has not proved easy to say what kind of creature made these footballs. They could have been the embryos of creatures akin to jellyfish and worms. Or they could have been the embryos of mysterious vendobionts like those fossils we met at Mistaken Point.

Embryos have for long excited thinking about evolution, and certainly long before 1859. Thus Darwin's own grandfather, the illustrious Lichfield doctor Erasmus Darwin, had mused upon the matter. And even Charles Darwin himself was very intrigued.[200] But various problems now present themselves to us with regard to this view about the Doushantuo 'embryos' being those of the earliest animals. As can be seen, we have entered dangerous territory—the arena for the mighty Mofaotyof game—of My Oldest Fossils Are Older Than Your Oldest Fossils. Needless to say, all the various Mofaotyof claims for the Doushantuo 'embryos' have been cautioned with health warnings, and even questioned, by the views of others. Which is just as it should be. Attackers in this game of palaeontological football say something like the following. First, no convincing adult forms have as yet been found. Second, the morphologies of the Doushantuo embryos are not uniquely 'animal' at all. They might be 'embryos' but not the 'embryos of animals'. They could instead be the embryos of algal colonies. Or worse, they might even be the husks of sulphur bacteria. The latter are very simple and primitive bacteria now known to thrive in settings with little oxygen but lots of sulphur and phosphate. For example, a giant sulphur-bacterium called *Thiomargarita* also undergoes cell divisions that are somewhat like those of 'embryos' from Doushantuo.[201] If that proves to be true (and it is fiercely contested, of course) then these embryos will tell us very little about the presence or absence of animals during the long dark age of the Precambrian.

Crucial in this game of football is the significance of dark little blobs found within just a few of the 'embryo' cells from

Doushantuo. These have been compared by some authors with dark little blobs found within eukaryote cells—each called a nucleus. This has then, in turn, been taken to confirm their animal embryo status. The problem with this line of reasoning is that even the simplest bacterial cells—which have never had a nucleus—can decompose to leave dark little blobs in their centres. Clearly, there is a danger of projecting on to the early fossil record, those things that we are hoping to find. As we have seen before, the proper question here is not: 'What do these blobs remind us of?' but 'What are these blobs in reality?' Or to put it even more crudely: 'Why can't a blob just be a *blob*?' Unhappily, the most boring explanation has an annoyingly high probability of being correct. That is what makes it a boring explanation.

It is here that the 'embryos' within Darwin's azoic phosphatic nodules—the Torridonian blobs of old Lake Beatha—come to the fore. Interestingly, our Torridonian 'embryos' from the Old Crone's Head are twice as old as those 'embryos' from Doushantuo in China—about 500 million years older. Even so, they share many features with them. The Scottish fossils show football-like clusters of cells in ones, twos, fours, and so on; footballs coated with 'egg shells' or resting cysts; cells with a dark nucleus-like blob inside them; and sheets and clusters of seaweed-like colonies. Interestingly, too, the Scottish and Chinese examples could both have been laid down in very shallow waters, from lakes, lagoons, or marginal marine settings. But whatever one chooses to make of these strange Precambrian cells and their owners, the potential of early phosphate to preserve ancient organisms at the cellular and subcellular level remains a truly astonishing facet of the early fossil record.

Haven't we heard of something like all of this before? Well, yes indeed we have. The same paradox—of better preservation in older rocks—was also met with in our study of preservation in the Ediacaran period. Strangely, again, such preservation is rarely seen by the early Ordovician, about 480 million years ago, and barely at all afterwards. Indeed, a comparable pattern can now be traced in nearly all kinds of preservation, including that of the carbonaceous Chengjiang and Burgess Shale biotas and the silicified Gunflint Chert type biotas. Pre-Cambrian preservation can be good, and sometimes brilliant. But post-Cambrian preservation is more often poor.[202]

So, here we have a paradox. The arguments from Lyell and Darwin predicted that the fossil record would turn out to read as follows. The fossil record is 'just a bit rubbish'. The Cambrian explosion is false—it was not an explosion of animal body plans. Those animal body plans may therefore have evolved a good deal earlier. In other words, the Cambrian explosion was little better than an explosion of fossils.

But, as we can now see, Darwin was more than a little confused by the nature of the early fossil record, or rather by the seeming lack of it. Signs of this confusion are scattered through relevant chapters in the *Origin of Species.*[203] But this confusion is forgivable and can now be resolved. Darwin had muddled up an essential distinction that needs to be made here—the difference between quantity and quality. Quantity in terms of fossils is not at all the same as quality in terms of the fossil record. And it is the *quality* of the early fossil record that is now starting to look a bit special. The evidence coming back to us from the fossil record, after a lifetime of study, is now starting to read like this. The pattern seems to be almost entirely upside down from what we had expected. Preservational quality seems to get better in rocks as we travel further back in time. This is broadly the opposite of the record anticipated by Charles Darwin in 1859. He had anticipated that preservation would become worse in older rocks. But to the contrary, the early fossil record is not really rubbish at all. It's just that the fossils tend to be small and cannot be studied without a microscope.

Not only have fossils clearly evolved through the course of time. The preservation of fossils has also 'evolved'. And most especially so during the Cambrian. That would imply that the Cambrian explosion was a real event. A Big Bang in evolution. An explosion of animals, not just of fossils.

From quality to quantity

Let me expand a little upon this inferred shift in the quality of the fossil record, from cells-to-dung, by taking us on a quick inspection of some curious patterns. As we have seen, phosphatic nodules from the Torridonian—about a billion years old—show clear evidence for exquisite preservation of cell walls and potentially for sub-cellular architecture. This pattern is also true for the famous microfossils from Doushantuo in South China—about 580 million years old and just a little older than the Ediacara biota. Here again, those fossils are superbly preserved by phosphate, revealing the shapes of mysterious 'embryos' and abundant algal colonies.

By the start of the Cambrian, however, remarkable preservation seems to be more difficult to detect. This is especially curious given the vast abundance of phosphatic deposits at this time and the huge numbers of rock slices that many of us have examined over decades of research. And it gets worse. After the Cambrian, it seems that such remarkable preservation of algal tissues and embryos has all but disappeared from the fossil record. In its stead are found rather shapeless agglomerations of bacterial crud. By the time we get to look at the Jurassic coprolites of old William

Buckland, or the phosphatic chalk of Thomas Henry Huxley, our rock slices contain little more than moderately preserved dung. This transformation mainly took place during the Cambrian period, roughly 540 to 480 million years ago. And, curiously enough, no one seems to have thought it worthy of much notice. But it matters very much indeed.

No less remarkable is a change that took place in the kinds of organisms that were themselves preserved. As we have seen, the earliest phosphate nodules from the Torridonian and from Doushantuo contain fossil assemblages dominated by seaweeds that grew under water in the presence of sunlight. In other words, they were likely to have lived in waters as little as a few tens of metres deep. By the start of the Cambrian about 540 million years ago, however, the preservation of algal cells in phosphate was starting to become less common. From that time onward, cellular preservation was increasingly confined to bundles of cyanobacteria-like sheaths and filaments, plus a few embryos and so-called 'jelly babies'. Even so, the water depths in which these earliest Cambrian phosphate nodules grew was still very shallow and probably sunlit. By the end of the Cambrian however, some 480 million years ago, even the cyanobacterial filaments were being squished together into shapes that look just like partly digested meals—plain old animal dung.[204] Gone forever by then were the phosphate-replaced seaweeds. That may be because the water depths in which we tend to find younger phosphate seem to have increased markedly, from tens of metres towards hundreds of metres.

No Precambrian trilobites?

The potential for such spectacular preservation of fossils in the Precambrian, such as that in phosphatic nodules, has interesting evolutionary implications, as well. It should allow palaeontologists to gain a much a better picture of the patterns taking place in the early evolution of the biosphere. At a certain point in our understanding, too, our ever-improving evidence from the early fossil record should allow us to test between competing evolutionary hypotheses, such as those we have mentioned earlier.

One of these competing ideas was called 'Lyell's Hunch'. That hypothesis, remember, suggested that the Cambrian explosion was not real, and that the Precambrian teemed with animals as yet unfound. Lyell's Hunch was the hunch originally employed by Charles Darwin to account for a lack of trilobites—or their ancestors—in rocks before the Cambrian. A fine example of Lyell's Hunch can still be found alive and well today. This version arises from the suggestion by highly respected scientists, such as Richard Fortey, that the missing animal ancestors were hiding somewhere on the Precambrian seafloor, perhaps living within the tiny pore spaces between sand grains. That would mean that they were very small indeed, perhaps less than a millimetre or so in length, and hence comparable with modern examples of the so-called meiofauna. Dredgings of any garden pond, when viewed down the microscope, will quickly show up a wonderful menagerie of meiofauna—multicellular animals so tiny that thousands of them could sit side-by-side on the head of a pin, if they could be circus-trained in that way. Here we encounter those beautiful little beasts called rotifers, nematodes, and gastrotrichs. Their tiny size is part of their strategy for survival in harsh conditions—low levels of oxygen, uncertain supplies of food. Conveniently, too, this meiofaunal size range is *exactly* the size range of fossils that we should expect to encounter as fossils in any of our ancient phosphatic nodules.

The test of any hypothesis lies, of course, in its predictions. Time, then, to put the predictive power of Lyell's Hunch to the test. Eminent scientists have spent their whole careers in pursuit of this. Mercifully for us, that means that many thousands of hours have been spent in putting this particular idea through its paces. One of our great hopes has been that of finding long-lost evidence for meiofaunal arthropods within Precambrian and Cambrian phosphate nodules. Interestingly, scientists have found exactly that: thousands of examples of microscopic arthropods exquisitely preserved in phosphate nodules from many parts of the world. But only from rocks of the Cambrian period. Nothing at all of this kind has yet been found in any Precambrian phosphate nodules. How very disappointing, you might say. But how very strange too because, as we have seen, the quality of preservation in phosphate before the Cambrian could be called 'peerless'. Yet micro-arthropods are as conspicuous by their absence in Precambrian nodules as they are by their presence in Cambrian ones.

How does all this chime with the timing of evolution in our Circus of Worms? It seems to match almost exactly. In rocks older than the Cambrian, there are *no* scratch marks left by the feet of arthropods—either big feet or small. And that can be regarded as a significant absence from the record because Ediacaran and older sediments were arguably almost as sensitive as photographic plates to all comings and goings on the seafloor—much more so than at later times.

So what are we to make of all this? First, we must admit that the *quantity* of the Precambrian fossil record can seem very slight indeed. There are no large animals, no bones, no teeth, and almost no skeletons. There are not even any convincing worm casts or jellyfish impressions. But what the Precambrian lacks in the bigness and quantity of fossils is surely compensated for in terms of its *quality* of fossil preservation. As we have seen, this high quality preservation has so far yielded no trace of arthropod animals, be they the remains of chitin skeletons, of legs, of faecal pellets, or even of scratch marks from their little hairy legs until the very start of the Cambrian period.

As we have seen with Lyell's Hunch, too, there comes a tipping point in science where negative evidence stops being an excuse and starts to become an important part of the evidence about pattern and process. That old Lyellian saying—absence of evidence is not evidence of absence—is vapid and rhetorical, of course. It is quite empty of any real meaning. Nor is it possible to say that negative evidence is any weaker than positive evidence. That is because negative evidence is an important part of the mathematical properties of pattern. Without negative evidence, science would find it impossible to make important medical diagnoses for cancer, to evaluate particle physics experiments, or DNA sequences, or molecular evolutionary studies. Even the printed words on this very page require 'absence'—the blank page—as well as presence—the ink itself. Likewise for word processing and the employment of zero and unity in binary computer codes. Thus it is with the pattern of evolution in the rocks—the evidence of evolution is as dependent upon negative evidence as is everything else in the real world.

Given the faithful preservation that phosphate can seemingly provide, this absence of arthropods and their relatives from the Precambrian fossil record—and most especially from our Precambrian phosphate nodules—may be telling us something important: *that trilobites and other arthropods did not emerge as a group until near the very start of the earliest Cambrian. So that arthropods, in the strictest sense, may well not have been denizens within Darwin's Lost World.*

So how may all these very marked changes in the *quality* of fossilization—near to the start of the Cambrian—be explained? This is not easy for us to answer because the variety of fossils affected by these changes seems to be so very great. It used to be argued that the Ediacaran fronds were preserved at Mistaken Point because they were provided with tough, leathery skins and were therefore more than usually 'fossilizable'. But, as we have seen, Ediacaran organisms were often rather thin and delicate. Nor do we find such a style of fossil preservation in marine rocks much younger than the Cambrian period, which is a bit hard to explain if they were so very tough and leathery. Likewise, the Doushantuo 'embryos' have been explained as having had a tougher and more 'fossilizable' envelope than that around embryos today. But, here again, this does not explain why we do not find 'embryos' in rocks much younger than the Cambrian.

When we find ourselves using a series of ad hoc explanations, then the explanatory power of those explanations might be said to be rather weak. What is needed in their place is a single unifying explanation that can be shown to have greater explanatory power. That is to say, we need to look for a single explanation in which all these various changes in preservation—from Ediacaran fossils and phosphatic 'embryos' on the one hand, towards Devil's Toenails and the Circus of Worms on the other—can be resolved by a single causal factor. Such an explanation should ideally explain, as well, those patterns observed in biosphere evolution, like those we have so far mentioned. If it does so, then it could be regarded as a more probable explanation than one that does not.

Darwin's vegetable patch

It will soon be time to put all our cards down the table, and to guess the name of the game. But just before we do so, it may be useful to make another visit to the home of Charles Darwin at Down in Kent. From 1837, Darwin had been casting around for evidence in support of his hunch that small causes can bring about very large outcomes. Lyell was already exploring the ways in which wind and water could slowly reduce a mountain range to sand. But Darwin was more interested in biological forces, akin to those needed for his emerging evolutionary ideas. Happily, he stumbled upon the world of worms and their wondrous impact upon the surface of the Earth: 'I was thus led to conclude that all the vegetable mould over the whole country has passed many times through and will again pass many times through, the intestinal canals of worms.'[205]

Near the end of his life, hamstrung by ill health and infirmity, Darwin had again decided to focus on something as near to home as he could possibly manage. Earthworms and their role in the formation of soil proved to be an ideal topic. He therefore began measuring them in his own vegetable patch, testing their rates of growth, their ability to turn over the soil each year, their ability to bury archaeological ruins, and even their ability to hear sound on the grand piano—alas, they could not enjoy his playing. Darwin concluded with these grandiose words: 'Worms have played a more important part in the history of the world than most persons would at first suppose... It may be doubted whether there are many other animals that have played so important a part in the history of the world, as have these lowly organized creatures.'[206]

His discoveries were so astonishing that worms became—for a time at least—the talk of polite Victorian society. They were found to both process and ventilate some fifteen tons of soil per acre per year. Worms took the wet and fallen leaves that cloaked the soil in autumn time, and carefully dragged them below the surface to line the burrow, where they could be digested at leisure. But they were also found to be cannibals, tearing at the chopped up remains of their colleagues. The health of the soil, on which so much of human society depended, was shown by Darwin to be somewhat out of our hands. Instead, it was down to the activities of little invertebrates like earthworms and earwigs, slugs, and slaters.

Humic soil formation didn't really get started on land until the middle of the Silurian period, some 150 million years after the Cambrian explosion. But there is, perhaps, such a thing as marine soil if we stretch our definitions a bit. We have seen something like it, for example, in the lagoons of Barbuda. In fact, we could argue that the Cambrian actually marked the beginning of soil-like formations on the seafloor.

A Whodunnit

Darwin, of course, was only able to guess a small part of the great game that we are now pondering. Those earthworms in his Victorian vegetable patch were only a small part of a much bigger picture. To appreciate the global importance of worms, we need to switch our attention from the soils of the world to the seas of the world. In these seas, we tend to take it for granted that the waters are healthfully ventilated from top to bottom by means of gaseous oxygen released by sea plants and land plants following photosynthesis. If that were not so, then the modern oceans would soon become stagnant, rather like a poorly maintained garden pond during the summer months. Toxic gases would tend to bubble up to the surface of such oceans, especially in places where the bottom waters well upwards, such as off the coasts of California and Peru today. Not such a bad thing, you might say. There would not only be lots of useful metals settling down into the muds, Los Angeles might also dissolve away forever under a great cloud of stink. But there would be many drawbacks, too. The stability of the Earth's climate would also be at great risk because oxygen, carbon, and climate are all intimately interlinked.

Not only that but oxygen in the atmosphere would dwindle. And animal brains are hungry for oxygen, so intelligence would be even harder to find than it is now.

It seems obvious enough that ventilation of the oceans is made possible today by the photosynthesis of land and sea plants. But it is a tad more complex than that. It is also made possible by the actions of multicellular animals. Both plants and animals can be visualized as working together like a well-organized Sanitary Corporation. Animals clean away the 'garbage bins' and tidy up the 'gutters' of the planetary surface. Back in the Cambrian, it seems that the new-fangled animals were also helping to clean up a kind of 'marine smog' that had previously tended to linger above the seafloor.[207] This cleaning they did, as animals still do today, by sweeping the water column clean of detritus, and most especially during the night shift, when zooplankton—such as copepods and krill—swim up to the surface waters to feed on the plankton. All that organic matter is then neatly packaged—along with clays and mineral matter—into droppings of the zooplankton. These tiny coprolites, once expelled, will tend to settle quickly and harmlessly downwards through the water column to land on the seafloor. By this noble and fastidious act, the waters of the whole ocean are cleansed on a daily basis, from top to bottom.[208]

But in a world before the evolution of animals and zooplankton—in the Precambrian ocean—conditions both below and above the seafloor must have been a bit like those in New York during a garbage strike. The seafloor, and even parts of the water column, will sometimes have resembled the toxic effluent that now comes oozing out of a smelting plant complete with its load of rust, lead, arsenic, and metallic sulphides. If time travel would allow us to go back to the late Precambrian on a Sunday afternoon, and then to start digging Darwin's underwater garden, we would surely have been overcome by the fumes of hydrogen sulphide and methane arising from the soil, unless we were wearing something like a space suit. The vegetables—mostly seaweeds and lichens perhaps—could well have been laced with cadmium and arsenic. And the worms and jellyfish—had they joined us in our quest—would have quickly dropped dead at our feet from the lack of sufficient oxygen. It would have been no use complaining, either, because those were the rules of the game.

The transformation brought about by the Cambrian Circus of Worms was therefore very dramatic indeed. Hydrogen sulphide and toxic metals—poisonous to all the higher forms of life—could no longer lurk just beneath the surface of sand and mud as they had for billions of years. Instead, the burrows of worms had arguably started to engineer something that resembled the air ducts of a mine shaft, wafting the sweet scent of oxygen down towards subterranean levels where the oxygen had barely ventured before. All of this is likely to have had a huge effect upon the chemistry and physics of the seafloor.[209] Not only that, but the calcium-rich shells of new-fangled skeletons were able to act together like giant global digestive tablets against the build-up of humic acids and toxic waste on the seafloor.

As for the toxic metals, so for the zone of phosphate formation. In Darwin's Lost World, before the Circus of Worms, phosphate was able to concentrate and then congeal upon the surface of the seafloor itself. It could act there like a honey trap, catching, and then embalming, the cells of nearby organisms before they could decay, as we have seen in the Torridonian and in Doushantuo. But once the Circus of Worms had evolved, near the very start of the Cambrian, the activities of animals on the seafloor is likely to have caused a rapid increase in the degree of mixing and churning of the sediment surface. Furthermore, the cells that lay near the surface were themselves consumed and mangled within the bellies of worms. The sediment became like a marine soil. All this processing was arguably beneficial because it helped the biosphere to recapture and to recycle valuable things like organic matter and phosphate ions. Animal activities also had the effect of pushing downwards the zone in which phosphate—and other minerals—could precipitate. These minerals now tended to concentrate and then congeal only at some distance below the surface of the seafloor. That meant that the stinking bacterial zones no longer lay at the surface but sat at some distance lower down. At these lower depths in the sediment, all manner of wonderful things that we now associate with the Precambrian fossil record—cells, embryos, seaweeds, and *Charnia*—had more time to 'go through the mill', passing through the bellies of worms perhaps several times before reaching the zone of phosphate at lower depths. They were therefore preserved as little better than dung (see Figure 22).

The magic of the Precambrian was no more. Darwin's Lost World had effectively been chewed to death.[210]

Achmelvich

While writing out these final thoughts, my little research group decided to return to Scotland, and the coastline around Loch Assynt and Lochinver. A small bay near here will always stick in my mind for its potential to help bring all this together—a place called Achmelvich. On the fine day when we arrived, the sea was tranquil. A small beach of bone-white sand was exposed at low tide, passing outwards into waters of almost mesmerizing clarity. Forests of brown kelp—each the size of a willow sapling—waved in the waters like shoals of drunken mermaids. Water and seaweed both seemingly beckoned us in.

Figure 22. A cartoon of the biological revolutions that took place in the 'topsoil' (approximately the top 10 cm) of seafloor after about 630 million years ago (at left), through the Ediacaran period (at centre) and into the Cambrian period, from about 540 million years ago (at right). Note the iceberg with falling dropstones at left, marking the Cryogenian ice age; and the emergence of skeletalized animals including tube worms, archaeocyath sponges and trilobites at right, marking the onset of the Cambrian period. The latter is also marked by deep and complex burrowing into the sediment. This 'circus of worms' and the onset of skeletons together transformed the chemistry of the seafloor. Before the Cambrian, rapid mineral growth on the seafloor could embalm cells and tissues rather well, like those of the Ediacara biota. After the Cambrian explosion, grazing and churning of the sediment began to close this remarkable window into the fossil record. In other words, the Precambrian fossil record is arguably much better and more reliable than Darwin would ever have dared to hope.

The rich clear waters of Achmelvich Bay are gently kissed by the Gulf Stream. These remnants of warm tropical waters caress the shoreline of a landscape sculpted out of ancient crags, themselves carved from crystalline Lewisian gneiss. To the north of the bay lie cliffs of russet and chocolate sandstone—beds left behind by Old Lake Beatha with its tiny treasures in phosphatic nodules and shale. On the eastern horizon can be seen the hills with the Pipe Rock.

Reflecting on all of this, I dropped down on to my knees to examine a handful of white beach sand with a lens. I could see myriads of tiny shells, just like those I had first met in the lagoons of Barbuda many years before—the remains of coralline algae, foraminifera, sponges, snails, and sea urchins. These felt like the echoes of that great Cambrian explosion which still reverberates around our shores. They were mementoes, too, to those heroes whom I found had crossed over into darkness while trying to tackle this huge puzzle—Hugh Miller, J. W. Salter, Captain FitzRoy, W. B. Carpenter, and Crosbie Matthews among them.

Here and there on the shore can be found cobbles of Torridonian sandstone—the remains of river beds over a billion years old. And in the cobbles lie shiny quartz pebbles that had once rolled and rattled along the floor of a riverbed long ago, after being eaten out of cliffs of jasper two billion years old. And in those jasper pebbles can be found microfossils from a world that is now two billion years past.

But that is a story that will have to wait for another time.

NOTES

1. See Darwin 1859: 308.
2. This is our modern image of Sir Richard Owen. See, for example, Desmond and Moore 1992. But see also Rupke 1994 for a case somewhat more in support of Owen's position.
3. See, for example, Desmond and Moore 1992 and Burkhardt 1996. Darwin's own references to Richard Owen also become more curt in successive editions of the *Origin of Species*. In the first edition of 1859, on p. 329, he is 'our great palaeontologist, Owen'. By the last edition of 1871, this falls back to 'Professor Owen' on p. 301.
4. Very readable explorations of Darwin and his world are given in the books by Janet Browne published in 2003.
5. For a clear and concise scientific summary of Darwinian evolution, see Mayr 2002.
6. Mendel was performing his experiments between about 1858 and 1863 but he did not publish until 1866. He made none of his findings public until a conference in 1865.
7. In truth, the palaeontologist Georges Cuvier was among the first to ponder on this, writing in 1812 that 'Life on this earth has often been troubled by terrible events . . . but what is even more surprising is that life itself has not always existed on the globe, and that it is easy for the observer to recognise the precise point where it has first left traces.' See Outram 1984: 156.

8. It is important to remember that Charles Darwin was an eminent geologist long before he was an eminent biologist. He had traversed 'Azoic' rocks in Wales with Adam Sedgwick as well as in South America on his own, during the 1830s. For Darwin as a geologist, see Herbert 2005.
9. See Darwin 1872: 286.
10. In 1859, what we now call the Precambrian–Cambrian boundary was usually regarded as the 'Primordial–Silurian' or 'Azoic–Silurian' boundary. For more on this early history, see Secord 1986.
11. I had the privilege of helping convene—with Michael House—the first ever international symposium on the Cambrian explosion and the origins of the major invertebrate groups, in April 1978. For an early in-depth overview of this event, see Brasier 1979 and House 1979.
12. According to the conventions of the time, Darwin followed Sir Roderick Murchison in placing the oldest known fossils in the Silurian Period. After a long and acrimonious dispute, the same rocks were later assigned to Sedgwick's Cambrian Period. This dispute is explored by Secord 1986.
13. Darwin 1859: 306. It is a matter of some interest to note the changes that took place in later editions, such as the following: 'There is another and allied difficulty, which is much graver. I allude to the manner in which numbers of species of the same group, [revised to 'species belonging to several of the main divisions of the animal kingdom' by the later 6th edition of 1872: 285–6] suddenly appear in the lowest known fossiliferous rocks . . . I cannot doubt that all the Silurian [revised to 'Cambrian and Silurian' in 6th edition: 286] trilobites have descended from some one crustacean, which must have lived long before the Silurian [revised to 'Cambrian'] age, and which probably differed greatly from any known animal.'
14. Darwin 1859: 307. By the 6th edition of 1872: 286, we note that Darwin had not only changed 'Silurian' to 'Cambrian', but 'my theory' had become 'the theory'. And his mention of 'quite unknown periods of time' had been removed.
15. See Conan Doyle, 1912.

16. The Royal Navy has a long tradition of natural scientists aboard its ships, including Joseph Banks in 1768–71, and Charles Darwin in 1831–6, not to mention those of the ill-fated Franklin Expedition to the Arctic in 1845–8. During the year-long voyage in 1970, I likewise shared my quarters with the officers, assisted with the making of charts and depth soundings, and took part in ceremonies—both official and unofficial.
17. The Hydrographic Division of the Royal Navy arose after Captain Cook's expeditions of 1768 onward, and was formally established in 1795. Much of our planet now carries the names of these great naval explorers and their ships: the Back River, Barrow Strait, Bass Strait, Beaufort Sea, Mount Erebus; and so the list rolls on.
18. Our aim during the cruise of *HMS Fox* and *HMS Fawn* was, in part, to help document the reefs and lagoons before commercial development would place it all under threat, as indeed, it later did. And before a new Panama canal directly connecting Pacific and Atlantic, would place Caribbean sea life under threat, which it did not because that canal was thankfully never built. The reefs and lagoons of Barbuda described in this chapter are now part of The Palaster Reef Marine Park, legally established by the Antigua Government in 1973. Unfortunately, all of Antigua and Barbuda's reefs are now seriously under threat from human activities. Both the coral reefs and the fish are in steep decline. The most pervasive threat is that of over-fishing. The chapter describes how these reefs and lagoons seemed to us during our early attempts at reefal monitoring back in 1970.
19. My scientific companions included Tom Barnard, Alec Smith, John Scott, Peter Wigley, John Wright, David Stoddart, and Peter Gibbs at various stages of the work.
20. Most of this work was followed up in a laboratory set-up within an old wartime Nissen Hut hidden in the bowels of University College London. This hut sat, somewhat auspiciously, beside the famous Flinders-Petrie Museum of Egyptology, and at the bottom of what was once Charles Darwin's old back garden in Gower Street. This is where he lived with Emma on his return from the cruise of *HMS Beagle*, from 1839 to 1842. It was in this garden that he developed his lifelong habit of a daily perambulation.

21. An outline of the modern tree of life is given by Guillaume and Le Guyader 2006, and authors in Briggs and Crowther 2001.
22. Strictly speaking symbiosis simply means 'living together', which can also include parasitism. The coral–dinoflagellate symbiosis is an example of a mutually beneficial 'mutualism'.
23. Darwin 1794.
24. Four distinct body plans had been envisioned by Cuvier back in 1812, namely the 'radiates' (corals, jellyfish and echinoderms), molluscs (clams, snails and cuttlefish) 'articulates' (crustaceans, spiders, insects and worms) and vertebrates (from fish to ourselves; see Outram 1984, Rupke 1994). The animal nature of sponges was for long a matter of debate since they do not move and have no organs. Feeding currents had been shown in sponges by Darwin's teacher Robert Grant, though Richard Owen was either antagonistic or unconvinced and therefore conspicuously excluded them from his Hunterian lectures on invertebrate animals. See, for example, Owen 1855. The lack of a head in echinoderms also resulted in their receiving a more lowly status than their current position as a sister group to chordates. Although Cuvier thought that these four archetypal body plans had never been bridged, the French evolutionist Etienne Geoffroy believed that there may have been links between them. For more on this, see Rupke 1994.
25. Cuvier had argued for the fixity of animal species and higher groupings such as families and classes. New species were 'created' after mass extinctions at the end of each major geological period. This led towards the tacit implication of a fresh creation at the start of every new geological period.
26. Lamarck did not believe in multiple episodes of creation. Instead he suggested that the habits and 'progress' made by one generation could be 'acquired'—passed on to the next generation.
27. Owen was often called the 'English Cuvier'. But Owen's own views changed from the position of Cuvier during the 1850s, towards a kind of divinely-directed evolution (see Rupke 1994). Darwin, of course, saw no evidence to support this idea of multiple episodes of creation. His hypothesis involved the higgledy-piggledy evolution of the Great Tree of Life, developed from a single ancestor in the Precambrian. For him,

these changes were extremely slow and controlled by the process of natural selection. Like Lyell, he questioned the evidence for mass extinctions at that time, though at least seven of these are now well-attested.

28. Though this is still very much under debate. See Cavalier Smith et al. 2006.
29. See Briggs and Crowther 2001, Guillaume and Le Guyader 2006.
30. Although the use of molecular clocks for divining Precambrian origins could be regarded as an example of weak inductive logic, it has much to teach us.
31. This formed the thesis for one of my first papers. See Brasier 1975.
32. Roland was working on the first ecological study of the Ediacara biota. See Goldring and Curnow 1967.
33. *Le Galerie d'Anatomie comparée et Paleontologie* is well worth a visit while in Paris because it conveys, for many, the intellectual spirit of the French Revolution and Napoleonic France. Here one can imbibe the astonishment French scientists felt about the similarities between ourselves and other living creatures. And all within a very beautiful building.
34. Baron Georges Cuvier lived from 1769 to 1832. See Outram 1984.
35. Some might think it odd that the presence of pores and canals was not considered enough to prove archaeocyaths to be sponges. But they lack spicules, which had always been thought a fundamental character of sponges. Some, such as Dorothy Hill of Queensland in 1972, therefore placed them within their own phylum. Others, such as Jack Sepkoski of Chicago thought they evolved too rapidly to have been sponges and speculated as to whether they were calcareous algae. But Rachel Wood at Cambridge and others made the important discovery that many fossil and living sponges have built a skeleton of dense calcium carbonate—including the stromatoporoids of the Devonian and the Sphinctozoans of the Cretaceous. The archaeocyaths were just the first of many such experiments of that kind.
36. The Precambrian–Cambrian boundary is the biggest geological datum in the rock record and I was honoured to be the Chairman in charge of the problem at the time of its ratification in 1992. But huge plaudits are

due to the energy, acumen, and oversight of Dr John Cowie of Bristol over several decades. For summaries of this huge international effort, see Cowie and Brasier 1989, and the authors in Lipps and Signor 1992.

37. For an entertaining account of what we have learned from trilobite evolution, see Fortey 2000. For a good general account account of palaeontology in this and following chapters, see authors in Briggs and Crowther 2001.
38. Darwin 1859: 308.
39. The classic 'Tommotian Bible' is that by Rozanov et al. 1969. For a later account in English about the Siberian sections, see Alexei Rozanov writing in Lipps and Signor 1992.
40. The Palaeontological Institute is a part of the Russian Academy of Sciences that is open to the public. Not only is the architecture very fine, the museum houses some of the best dinosaur fossils from the Gobi desert as well as frozen mammoths from Siberia with parts of their DNA preserved.
41. I was among the first to succumb to this fabled epidemic of *Giardia*, that swept through our camp and hamstrung much of our later work on board the *SS Rossiya* along the Lena River. On one dreadful morning at 3 a.m., there were eleven of us queuing in pyjamas in the hope of access to a single toilet. Leaving the ship in the dock, I remember walking away against a backdrop of sound—the echoes of colleagues still vomiting in the bowels of the ship. Our dysentry gave a darker meaning to the term 'Cambrian Explosion', which became the ribald name for those 'explosions' of the alimentary canal that we experienced on this legendary trip. It took some of us several years to get back to strength.
42. Or Barry Manilow or Sacha Distel of evolution, if that makes more sense.
43. The great Cambrian palaeobiologist Charles Walcott once thought he had found brachiopods in the Precambrian of the Grand Canyon. See Walcott 1899. Unfortunately, his fossils of *Chuaria, Beltina*, and related forms proved controversial and are now thought little better than rather featureless bacterial envelopes or algal cysts. They are without

evidence for cellular structure, as shown by the late Dr Gonzalo Vidal and others.

44. Small shelly fossils were to dominate thinking about the origins of the Cambrian explosion for the decade from 1978 in Cambridge where we decided they were of crucial importance, to 1987, when the signs of animal activity called 'trace fossils' took precedence. Absolutely key in this work was an international meeting at Uppsala in Sweden in 1986, organized by Stefan Bengtson. Unhappily it coincided with the fallout from the Chernobyl atomic reactor explosion but it brought all workers on the Cambrian explosion, and their fossils, together under one roof. Key papers on small shelly fossils include that by Qian and Bengtson 1989, and those summarized by authors in Cowie and Brasier 1989, and in Lipps and Signor 1992.
45. See Hou et al. 2004 for an excellent atlas of these fossils and Gabott et al. 2004, for an account of their preservation.
46. For a philosophical take on the Burgess Shale fossils, see Gould 1989 as well as its intended antidote, Conway Morris 1998. For a student text and key papers, see Selden and Nudds 2004. For a fine atlas of these fossils, see Briggs et al. 1994.
47. Darwin 1859: 306 and similar on p. 313.
48. Darwin 1859: 316.
49. Darwin 1859: 338.
50. Studies of growth lines in brachiopod shells from Chengjiang are now showing us that they also comprised generational cohorts that grew up and died together.
51. See Gould 1989.
52. This range of body plans was called 'disparity' by Gould. It is important to note that disparity is not the same as diversity. It is like the reading of a hand of cards—diversity is a measure of the numbers of different cards, say a King, ten, eight, five, and two of clubs; whereas disparity is a measure of the extremeness of the end-members, like a King and a two of clubs. One hand may have the same diversity but a much greater or smaller disparity.
53. For more on this spirited debate, see Bill Bryson 2003. For some of the key protagonists, see Gould 1989, and Conway Morris 1998 and 2003.

54. The Novosibirsk view of Missarzhevsky and others was distinct from that in Moscow. They seemingly regarded the Chinese fossils as of broadly the same age as their (pre-Tommotian) Nemakit–Daldynian Stage.
55. The Siberian sections had other challenges too, including their inaccessibility. Eventually, that was to be their undoing as the global reference section for the Precambrian–Cambrian boundary; see Cowie and Brasier 1989. Even so, they have many key virtues, including their potential for chemical studies and their relative completeness; as shown by Brasier, Rozanov, et al. 1994.
56. I was a guest of Dr Xiang Liwen of the Museum of the Ministry of Geology in Beijing. We were planning to produce a book on the Cambrian explosion in China. After three years of work on maps, sections, and text, the British publisher, Scottish Academic Press went bankrupt and my Geology Department at Hull was closed down owing to cuts in funding under the Thatcher government. I therefore moved to Oxford and that project has slept in a store room ever since.
57. The Great Proletarian Cultural Revolution was a result of the struggle for power with the Communist Party of China. Launched by Chairman Mao Zedong, it lasted from 1966 to 1976.
58. This museum, called the Geological Museum of China, lies in the now flourishing Xisi area of Beijing and is the largest such museum in Asia. It opened in 1959 and houses the giant Shandong dinosaurs, the early birds of Laioning, and, of course, the remains of 'Peking Man'.
59. For a general account about the earliest Cambrian small shelly fossils of China, see Jiang Zhiwen writing in Lipps and Signor 1992.
60. Darwin 1859: 288.
61. It later proved possible for us to trace this and similar layers beyond the Himalayas into Oman by the Arabian Gulf, from there into Iran south of the Caspian Sea, and thence into Spain in the Mountains of Toledo. See the chapter on this by Brasier writing in Cowie and Brasier 1989.
62. See Bengtson and Zhao 1997.
63. See Darwin 1859: 307. Murchison mostly kept his views on progressive creation to himself. But he wrote the following interesting letter to Professor Harkness in about 1860: 'if you read the work of Darwin on

the Origin of Species, which has given an earthquake shock, you will easily see that in reality, my geological postulates, if not upset, destroy his whole theory. He will have no creation—no signs of a beginning—millions of living things before the lowest Silurian [Cambrian]—no succession of creatures from lower to higher, but a mere transmutation from a monad [a single-celled organism] to a man. His assumption of the position of the Lyellian theory, that causation never was more intense than it is now, and that former great disruptions (faults) were all removed by the denudation of ages, is so gratuitous, and so entirely antagonistic to my creed, that I deny all his inductions, and am still as firm a believer as ever that a monkey and a man are distinct species, and not connected by any links, i.e. are distinct creations. The [Lyellian] believers in a lower, and a lower still [without any progress], have never answered, and cannot answer, the fact that the rich marine Lower Silurian fauna is invertebrate, and cannot answer the fact that the Cambrian rocks of Ireland, Wales, Shropshire, and the northwest of Scotland, though less altered than the Lower Silurian, have afforded nothing distinct which is higher than an Oldhamia or a worm.' See Morton, 2004: 196–7.

64. See Lyell 1853: 134.
65. For more on the Coelacanth, see Weinberg 1999.
66. See, for example, Darwin 1861: 327 who inserted this into his third edition: 'In all cases positive palaeontological evidence may be implicitly trusted; negative evidence is worthless, as experience has so often shown'.
67. In 1837, Lyell wrote: 'We must not, however, too hastily infer from the absence of fossil bones of Mammalia in the older rocks, that the highest class of vertebrated animals did not exist in the remoter ages'. See Lyell 1837: 232. Part of the problem, here, was that Palaeozoic fossils had been incorrectly compared with rather modern groups like palm trees. By the 1850 edition of the *Principles*, on p. 135, a note of caution had started to creep in: 'Endeavouring, however, to ascribe a due share of importance even to negative testimony, we may regard the non-discovery hitherto of fossil cetacea [whales] in all the rocks from the Silurian to the chalk inclusive, as offering the most striking fact in

Paleontology in favour of the doctrine that the most highly organised class of animals was one of the last which made its appearance on the earth'. But yet again, by 1853, on p. 137, all such hostages to fortune had been deleted and a whole new section was added on the problems of the fossil cetaceans, ending with this spectacular example of British understatement: 'In the present imperfect state then of our information, we can scarcely say more than that the cetacea seem to have been scarce in the secondary [Mesozoic] and primary [Palaeozoic] periods.' Lyell was seemingly baffled by the whole matter until Darwin helped him to make up his mind. In the *Elements* of 1865: 586, we find him saying: 'as we trace [fossil groups] farther and farther back into the past, it cannot be denied that our failure to detect signs of them in older strata, in proportion to the rank of organization, favours the doctrine of development, or at least of the successive appearance on the earth of beings, more and more highly organized, culminating at last in the appearance of man himself.' By 1871, Lyell was to write, in the *Elements* on p. 375: 'The predominance in earlier ages of these mammals of a low grade, and the absence, so far as our investigations have yet gone, of species of higher organization, whether aquatic or terrestrial, is certainly in favour of the theory of progressive development.'

68. In other words, evolution predicts that the fossil record should preserve a story of 'progress' within distinct lineages over millions of years, as indeed it does. See Michael Le Page, *New Scientist*, 19 April 2008: 26.
69. See McIntyre and McKirdy 1997 and Hallam 1983.
70. In truth, this idea goes back to Leonardo da Vinci and Nicolaus Steno. See for example, Cutler 2004, and Kemp 2006. But it was arguably Werner who first attempted to test a hypothetical process by means of the concerted study of patterns in the rocks.
71. See Chambers 1844, who estimated the depth of the primary ocean to have been about 100 miles deep, presumably in order to account for the metamorphism of the crystalline 'Primary' rocks. If we make some allowance for their depth within the crust, rather than depth within the ocean, he was not altogether wide of the mark. Recent estimates of early ocean temperatures have come up with average values of about

70°C according to Knauth 2005. Interestingly, even Darwin introduced a related remark into his 1861 edition, saying 'I may remark that all geologists, excepting the few who believe they see in the metamorphic schists and plutonic rocks the heated primordial nucleus of the globe, will probably admit that rocks of this nature must have been largely denuded. For it is scarcely possible that these rocks should have been solidified and crystallized in a naked condition; but if the metamorphic action occurred at profound depths of the ocean, the former mantle may not have been thick'; see Darwin 1861, 3rd edn: 314-15. While Darwin was writing the *Origin of Species*, many senior scientists, including Professor Sedgwick of Cambridge, thought that evidence from the fossil record—that is to say—no life, followed successively by corals, shellfish, fish, reptiles, mammals and man—reflected not the effects of evolution so much as the results of a progressive cooling of the planetary surface. In this view, the lack of life before the Cambrian was due to infernal heat. Corals and shellfish were able to appear once temperatures had dropped to form the hot primordial ocean. Mammals and man then came to occupy an Earth made progressively more temperate by 'the Creator'. See Rupke 1994, for a discussion of this theme.

72. See Daly 1907.
73. There is rather good evidence for the concept of a primordial ocean with a strangely different chemistry from that found today. It seemingly had much less oxygen and much more iron. See, for example, authors in Cavalier-Smith et al. 2006 and Leach et al. 2006.
74. Here is how Sir Edgeworth David put this in 1936: 'the existence of life for aeons before Cambrian time has been inferred from the great differentiation of animal life even in the Lower Cambrian rocks. Palaeontologists claim that even the earliest known forms of Cambrian life are already a long way up the life-column. Next comes the consideration as to what evidence might reasonably be expected in pre-Cambrian rocks in regard to the nature of contemporaneous life, and in what form the fossils are likely to be preserved. A study of mineral composition of fossils, from late Palaeozoic back to early Palaeozoic time, shows a dwindling of thick-shelled calcareous types, such as the Pelecypods

[clams], whereas Brachiopods like Lingula . . . with their thin shells, formed of horny material and phosphate with carbonate of lime in alternate layers, become relatively more numerous as we move back in time. With the exception of the Archaeocyathinae, and certain calcareous Algae, this rule that the amount of calcite [chalky material] in the hard parts of the animals lessens with increased geological age holds good down to the base of the Cambrian, animals of earlier age may be assumed to have been either soft-bodied or chitinous, or silicon-secreting forms. Such soft-bodied animals probably abounded in the pre-Cambrian seas, but apart from their trails or the casts of coelenterate body-cavities, as in forms allied to Medusites, these would leave behind them little trace of their former presence.' See David and Tillyard 1936.

75. See Sollas 1905: 28. First delivered at Bradford in 1900.
76. The Piltdown Hoaxer brought together an orang-utan jaw and a human cranium and passed them off as the remains of the earliest Englishman, in about 1912. The main perpetrator seems to have been a local solicitor called Charles Dawson. But Oxford Professor Archibald Douglas argued that his old boss, William Sollas, was the real brains behind it all. This claim now receives little support, in part because there is evidence that Sollas was never near to the crime scene. See, for example, Spencer 1990.
77. This Mongolian expedition of the International Geological Correlation Program, Project 303, had the following team members: Dorj Dorjnamaa and Y. Bat-Ireedui of Mongolia; Vsevelod Khomentovsky from Russia; myself, Roland Goldring, Rachel Wood, and Simon Conway Morris from England; Françoise and Max Debrenne, and Pierre Courjault-Rade from France; Anna Gandin from Italy; Ken Hsu and Graham Shields from Switzerland; John Lindsay, Pierre and Peta Kruse from Australia. A whole host of others were drawn in for the full report, including Joe Kirschvink, Dave Evans, A. S. Gibsher, and Soren Jensen.
78. A short account of the stratigraphy and palaeontology is given by the author, in Cowie and Brasier 1989.
79. Endnote, Darwin 1861, 3rd edition: 315.

80. None of this would have been possible without the many years of hard work undertaken in Mongolia by Russian palaeontologists such as Nadir Esakova, Alexei Rozanov, Galia Ushatinskaya, Lena Zhegallo, and Andrei Zhuravlev in the years before our second expedition. Most of this work is published in Russian, see Voronin et al. 1982. An expansion of this work was published in an issue of *Geological Magazine*, see Brasier et al. 1996 and 1997.
81. See Raup 1966.
82. Darwin 1859: 288.
83. Darwin 1861: 321, wrestled with the problem that older and younger taxa do not always appear in neat order. The details of this lineage are still being worked out.
84. Darwin 1859: 302–3.
85. See Darwin 1859: 317.
86. See Parker 2003.
87. Molecular phylogenies mostly confirm a primitive position for the Arrow Worms, which are formally placed in the Phylum Chaetognatha. See Marletaz et al. 2006.
88. See Bak 1997: 77. That is a range over eleven orders of magnitude. Such a logarithmic pattern is called a Power Law distribution.
89. This mountainous metaphor, of the Cambrian Cascade, has its limitations of course. But it does explain some of the curious phenomena that are shared between the Cambrian explosion and the phenomenon of landslides. Like landslides, large evolutionary phenomena—such as the formation of phyla—are rare whereas small phenomena—such as the formation of species—are common. Indeed, when the frequency of taxonomic layers within a given animal group, or within the whole animal kingdom, is tallied, it shows something like the Power Law pattern seen in landslides and earthquakes. That is a common feature in all historic phenomena, of course. The Cambrian Cascade analogy also makes some interesting predictions that have yet to be tested. The first appearance of the animal phyla may have been preceded by a biosphere system that was 'going critical'. One possibility is that the evolution of multicellularity in the Proterozoic—without ecological tiering—was otherwise leading towards biogechemical cycles that were

inherently unstable and associated with the so-called Snowball glaciations. The splitting-off of large taxonomic groups, such as phyla, may have been preceded by 'foreshocks of divergence' and followed by 'aftershocks'. The causes for the origins of the animal phyla may not be distinguishable from the causes for the origins of species. Divergences leading to the origins of new phyla may be explained by the chance alignment of multiple disconnections from the parent body—such as Hox gene mutations, as well as geographical and seasonal isolations. These evolutionary avalanches clearly involved a symmetry-breaking cascade—in which radial symmetry, for example, developed into bilateral symmetries of the two main kinds, called protostomes and deuterostomes. Very large divergences—such as the origins of the animal phyla—may be seen as having taken place in 'swarms'—especially from the end of the Ediacaran to the Tommotian.

90. A classic example of inductive logic is the Hox gene explanation for the explosion of animals. It runs like this. Hox gene mutations *could* generate macroevolutionary changes. There were macroevolutionary changes. Therefore it was the Hox genes wot dunnit. But the premises do not entail the conclusion. Therefore, arguably, the conclusion is weak.
91. Much the same chain of events is likely to have taken place within the evolution of calcareous algae, to evolve towards forms ancestral to the *Halimeda* we met in Chapter 1.
92. Soviet leaders Andropov and Brezhnev thought that Britain and America were preparing to launch a pre-emptive nuclear strike against Russia, following the 'Star Wars' speech by US President Ronald Reagan early in 1983.
93. From *Alice Through The Looking Glass* by Lewis Caroll. For a whimsical philosophical gloss on that text, see Heath 1974.
94. See Secord, 1986 and Morton 2004, for more on Murchison and his career.
95. The first fossils ever described from Cambrian rocks were the trilobites *Entomostracites* (later *Agnostus) pisiformis* by Wahlenberg in 1821 and *Trilobitus* (later *Ellipsocephalus) hoffi* by Zenker in 1833, both from Czechoslovakia. Their great age was not demonstrated, however, until the work of Barrande.

96. Darwin 1859: 307.
97. Darwin 1872, 6th edn.: 287.
98. For a general introduction to trace fossils and the Cambrian explosion, see Crimes writing in Cowie and Brasier 1989, and Crimes writing in Lipps and Signor 1992.
99. Although *Callavia* is the earliest trilobite to be found across the Avalonian region, it was a little younger than *Fallotaspis* in Siberia and maybe of *Eoredlichia* in China. See Brasier in Cowie and Brasier 1989.
100. Avalonia takes its name from the Jacobean colony of Avalon, set up in south-eastern Newfoundland by Lord Baltimore in 1627. He used this name, no doubt, to encourage fond memories of the legend of King Arthur who had retreated to a mythical island in the west called Avalon—meaning the Place of Apples.
101. See Landing et al., 1988, Conway Morris writing in Cowie and Brasier 1989, as well as Ed Landing's chapter in Lipps and Signor 1992.
102. The coast of Fortune Head is a National Park of Canada. It was established as a provisional reserve in 1990 and then given full ecological reserve status in 1992 following its selection as the global stratotype for the Precambrian–Cambrian boundary (see Brasier et al. 1994). There is a fully staffed Visitor Centre, open during the summer months, to welcome tourists interested in exploring the Cambrian Explosion, complete with finely crafted dioramas.
103. Be warned. Screech is the local strong spirit made from raw Jamaican Rum. 'Screeching-in', as my students Rich Callow and Alex Liu soon discovered, typically includes the following charming rituals: being blindfolded; dressing up in fisherman's waterproofs and sou'westers; entering a mock boat; being pelted with water; eating fish eyes; swallowing 'hard tack'; drinking screech; and dancing around an oar, all to the cheering support of onlookers.
104. These 'Pre-Trilobite' rocks are now called 'Fortunian', and mark the earliest formally defined stage of the Cambrian. The Fortunian is of approximately the same age as the Nemakit-Daldynian in Siberia, and of the Meishucunian in China.

105. This fossil was at first called *Phycodes pedum* (see Brasier et al. 1994) until it became apparent that its behaviour was different from true *Phycodes* and the species was transferred to the genus *Trichophycus*.
106. See, for example, authors in Landing et al. 1988, and Brasier et al. 1994. Unhappily, this decision was a cruel blow to Russian pioneers such as Alexei Rozanov and Boris Sokolov, and to the new Chinese teams led by Xing Yusheng. But geological boundaries are not written in stone. Future generations may turn things around.
107. And once the Cambrian explosion had started, some say, the rest of the game—colonization of the land, trees, large quadrupeds, intelligence, even humans perhaps, were almost inevitable outcomes. For this almost 'Designist' view, see Conway Morris 2003. For the opposite 'Contingentist' view—that 'if we were to re-run the tape of life from the Cambrian Explosion, everything would turn out differently, see Gould 1989.
108. The Mistaken Point Ecological Reserve has been short-listed for UNESCO World Heritage Designation. There is now a Visitor's Centre at Portugal Cove South.
109. See Narbonne 2004. They are not truly fractal structures, of course, because they are not self similar at all scales.
110. See Ansted 1866: 47–8.
111. See W. W. Watts 1947. In all other respects, though, this is a groundbreaking book on the geological history and structure of Charnwood Forest.
112. I am very grateful to Roger Mason, Trevor Ford, and Helen Boynton for sharing their reminiscences on all this over the years.
113. Professor Peter Sylvester-Bradley of Leicester was also drawn into this work. But note that it took a succession of *amateur* geologists to make professional geologists accept this most astounding fact—that large fossil remains can be well-displayed in rocks a good deal older than the Cambrian. See Ford 1958 and 2007.
114. Fortnum & Masons is a famously snooty Regency department store that sells exotic foods—such as pickled quails' eggs, or ants embalmed in chocolate. It also sits exactly across the road from the equally grand chambers in Burlington House, where earth and life scientists still meet

to debate their discoveries. Darwin's first abstract for the *Origin of Species*, for instance was read out in the rooms of the Linnaean Society, on the opposite side of the road. And next door, the debate on the famous Piltdown hoax had taken place in the rooms of the Geological Society.

115. See Sprigg 1947 and 1949. For more recent introductions to the Ediacaran fossils, see Glaessner 1984, R. J. F. Jenkins in Lipps and Signor 1992, and Gehling in Fedonkin et al. 2007.
116. The Ediacaran Period is the first significant subdivision of geological time to have been erected in our lifetimes. The previous one, the Ordovician Period, was put forward by Charles Lapworth in 1879. The present author was honoured to be on the Voting Panel for this new period, which was ratified in March 2004. The final hard graft was undertaken by authors in Knoll et al. 2004.
117. For a light-hearted read on the recollections of a mineral geologist, see Sprigg 1989.
118. I say 'much', here, because he was also a world expert on microfossil evolution, on the evolution of fossil crabs, and even on Maritime Law.
119. See Glaessner and Wade 1966, and Glaessner 1984.
120. This idea of continuity between the Ediacara and Cambrian biotas quickly gained acceptance, especially with the inception of the international Working Group on the Precambrian–Cambrian Boundary in 1972, after which it began to be argued that trilobites were 'Cambrian' whereas the Ediacara biota was 'Precambrian', with the boundary lying somewhere in between; see Harland 1974, Cowie and Brasier 1989. In this respect, it is important to remember that the base of the Cambrian had never been specified clearly by Professor Adam Sedgwick of Cambridge back in the mid-nineteenth century, leaving the concept wide open for future discussion. This Precambrian–Cambrian boundary was actually argued to lie *below* the Ediacara biota by Cloud and Nelson 1966. Its present position above the Ediacara biota was not internationally ratified until 1994, at the level of first appearance of the trace fossil assemblage containing *Trichophycus pedum*. See Brasier et al. 1994.

121. These Namibian fossils had first been described in a series of little noticed publications by Gürich in the 1920s and onward through the 1930s (see Gurich 1933). At that time they were thought to be of Cambrian age. For an illustrated review, see Pat Vickers Rich in Fedonkin et al. 2007.
122. See, for example, Pflüg 1972.
123. For statements of the vendobiont hypothesis, see Seilacher 1992, and Buss and Seilacher 1994.
124. Mark McMenamin has written several thought-provoking books on the Cambrian Explosion and the Ediacara biota. See, for example, McMenamin 1998.
125. The largest specimen in the Darwin Centre galleries that we examined was a giant squid, some six metres long.
126. See Antcliffe and Brasier 2008.
127. Darwin 1897: 111–12. I have used the copy that I took with me on the Voyage of *HMS Fawn*. This book was originally published in 1845.
128. See Brasier and Antcliffe 2004; Antcliffe and Brasier 2008.
129. This is widely agreed for *Fractofusus*. Dima Grazhdankin of Novosbirsk has accumulated evidence to suggest that even *Charnia* lived face-downward in, or within, the mud. See also Grazhdankin 2004.
130. See Fedonkin 1990. For Misha Fedonkin's influential thinking on the Russian White Sea biota, see Fedonkin in Lipps and Signor 1992 and in Fedonkin et al. 2007.
131. See Runnegar 1982.
132. The inkblot or Rorschach test for health patients involves being shown, for example, a pattern that resembles either a 'black wine glass on white background', or a 'white couple kissing against a black background' according to the state of mind.
133. See Gehling 1987.
134. There are numerous accounts of the Fibonacci series. See, for example, Ball 1999, Stewart 2001 and Brasier and Antcliffe 2004, written in iambic pentameter!
135. See, for example, Grazhdankin and Gerdes 2007.
136. Fedonkin and Waggoner 1997.
137. See Fedonkin et al. 2007.
138. See Salter 1856, 1857; Secord 1986.

139. Darwin 1859: 307.
140. See McIlroy et al. 2005.
141. These new finds found by Richard Callow and Alex Liu are three dimensional and preserved in clay minerals, which adds further to the report of carbonaceous filaments in thin section in the Longmynd rocks, first reported by Peat 1984.
142. See Twemlow 1868.
143. The early history of research into the earliest animal life is full of Mofaotyofs, many from the most esteemed pens—Mr J. W. Salter and *Palaeopyge* from the Longmynd; Dr W. B. Carpenter and *Eozoon* from Canada; Professor Cayeux and his 'sponge spicule' *Eospicula* from Brittany. These Mofaotyofs or other Mofaotyof creatures happily form the culmination of a book by Bill Schopf of Los Angeles, called 'The Cradle of Life'. See Schopf 1999.
144. Most notorious here is, of course, the mysterious case of the Piltdown Skull, reported in a series of papers to a gullible public by Charles Dawson and Arthur Smith Woodward. The latter was Keeper of Geology at the British Museum of Natural History at South Kensington in London. And he managed to limit the access of his rivals to the actual bones for little more than twenty minutes each or less, in their lifetimes. For the rest of the work, they had to work with casts he had made. That somewhat scandalous behaviour by Smith Woodward is arguably how and why this greatest ever scientific hoax managed to stay current for so long. See Spencer 1990.
145. For an entertaining book on the geology of malt whiskies from Islay and beyond, see Cribb and Cribb 1998.
146. See, for example, Jim Gehling in Fedonkin et al. 2007.
147. Strange as it may seem, the presence of chordates has been claimed in popular books and even before the media; see Gehling in Fedonkin et al. 2007. There have, as yet, been no dubious claims for dinosaurs or ape-men—merely a matter of time, perhaps.
148. The role of worms in the production of marine soils near the start of the Cambrian will be examined again in more detail in Chapter 9.

149. See, for example Brasier et al. 1997, and authors in Fedonkin et al. 2007.
150. Darwin 1859: 366.
151. See Cribb and Cribb 1998.
152. The Cryogenian is a new period that has yet to be formally defined.
153. See Hoffman and Schrag, 2002 for a recent technical account and the books by Walker 2004 and MacDougall 2006 for more popular accounts.
154. The national oil company of Oman is Petroleum Development Oman, which is affiliated to Shell Petroleum.
155. See Germs 1972 and Pat Vickers-Rich in Fedonkin et al. 2007 for Namibian examples of *Cloudina* and *Namacalathus*.
156. See Grotzinger et al. 2000.
157. For more on this golden spike, see Knoll et al. 2004.
158. This is named, of course, after the Monty Python sketch 'Mr Creosote', whose character was played by my Teddy Hall (St Edmund Hall) colleague, Terry Jones.
159. See, for example, Bak 1997.
160. Other things being equal, of course.
161. That is because their surface provides the area on which oxygen, or enzymes or other catalysts, are obliged to act. If the relative surface area is less, then the effective oxidation of the cell will be less. Other things being equal, of course.
162. And it means that they are less likely to be consumed and respired, and then returned to the atmosphere as carbon dioxide gas.
163. See, for example, Brasier in Lipps & Signor 1992, Brasier & Lindsay 1998; Brasier 2000.
164. Lyell 1871: 88.
165. In truth, I learned about those paradoxical trilobites from a visit I made in 1989, when I had to confirm where a mystery trilobite had really come from. This specimen of *Olenellus* was lodged in the Oxford University Museum, with a label saying it had been collected from the basal quartzite at Knockan Cliff. That would have made it the oldest known trilobite in the region. Fieldwork showed that it almost

certainly came from higher, in the Salterella Grit, and just below the Moine Thrust at Knockan Cliff.

166. See Peach, Horne, et al. 1907.
167. See Secord 1986.
168. Stromatolites, also of Cambrian age, from 'down the road' in New York State, were called *Cryptozoon* by the indomitable Charles Walcott in 1883, and regarded by him as reef-like constructions of algal colonies. Such structures had even been reported from the 2 billion-year-old Gunflint Chert of the USA and Canada by geologists J. D. Foster and J. W. Whitney as early as 1851. But it was not until the pioneering work of Cambridge geologist Maurice Black in the Bahamas of the 1930s that the role of cyanobacteria in their construction became a bit more clear.
169. Lyell 1971: 89. They were at first thought by him to be Silurian in the sense of Murchison, and not Cambrian as we now know them to be.
170. Hugh Miller 1858: 328–30.
171. The term Torridon Sandstone was introduced by Nicol; Torridonian was first introduced by Archibald Giekie. The rocks take their name from Loch Torridon.
172. Darwin 1859: 343. In the 6th edition of 1872, this was changed to read: 'long before the Cambrian system was deposited? We now know that at least one animal did then exist; but I can answer this question . . . ' Darwin was here presumably referring to *Eozoon*, which will be discussed later in this chapter.
173. See, for example, Temple 2000.
174. See David Hockney 2001.
175. For more on Robert Hooke, see Jardine 2004.
176. For Forbes, but also for Lyell; see his *Principles*, Lyell 1850: 134.
177. Darwin 1859: 291.
178. Darwin 1861, 3rd edn.: 312.
179. See Huxley 1894.
180. The correct nautical term is not 'underway' but 'under weigh', meaning that the anchor has been 'weighed' or lifted.
181. Lyell 1865: 579.

182. See Dawson 1888: 15–18. Note also that Dawson did not include *Eozoon* in this paragraph because he believed it to be the remains of an early animal, not a plant.
183. See, for example, Carpenter 1891.
184. This mapping trip took place during the run-up to the famous Apollo moon landing, and geological sampling, of 1969. Lunar geology was then as much the talk of town as Martian geology is today.
185. Darwin 1872: 287.
186. See Darwin 1872: 308.
187. In truth, the latter do not really belong in the family of science at all.
188. Darwin 1861, 3rd edn.: 306–7.
189. The Pilbara rocks are from 2700 to 3600 million years old. The story of their decoding, and of the origins of life itself, is now in preparation by the author.
190. Darwin 1859: 307. This was revised to read 'The presence of phosphatic nodules and bituminous matter even in some of the lowest azoic rocks probably indicates life at these periods', in the 1872 edn.: 287.
191. Teall gave the following first description of cells in the *Summary of Progress for the Geological Survey of the United Kingdom of 1902*, p. 56: 'The lenticles [of phosphate] consist of the finer portions of the sedimentary material, especially micas, cemented with amorphous phosphate. They contain also minute black spherical bodies, brown fibres, or black or brown shreds, some of which show distinct cellular structure. The cells measure about 0.01 m.m. in diameter and are clustered in groups. The natural form is spherical, but this is sometimes interfered with by mutual pressure so that the cells have polygonal outlines. In such cases the fragments resemble portions of the parenchymatous tissue of plants, except as regard size [which is smaller]. The black spheres show no structure where they are wholly immersed in the phosphatic matter; but when as happens at least in one case, about half has been removed in the process of making the section, they are then seen to have been hollow and apparently perforated. The spheres vary from about .005 to .035 m.m. in diameter . . . The brown fibres are about .004 m.m. in breadth and of length varying from a few hundredths to tenths of a millimetre. They may be

straight, curved, or even looped... That these various bodies represent organisms is highly probable; but the nature of the organisms is doubtful.' I am indebted to Leila Battison for pointing me toward this document.

192. See Peach et al., 1907.
193. Phosphorus bombs and phosgene gas were among its many terrible uses. See Emsley 2000.
194. This work began in 1953. See Barghoorn and Tyler 1965, Schopf 1999, and Knoll 2003.
195. Though the flag was kept waving in a short article by Chris Peat and Bill Diver in 1982.
196. Darwin 1859: 288.
197. See Darwin 1897: 338. Silex here is the old name for chert—microcrystalline silicon dioxide that can preserve cells with great fidelity.
198. Darwin 1897: 31.
199. For the animal embryo idea, see Knoll 2003; Xiao et al. 1998 and 1999; Donoghue et al. 2006.
200. See Darwin 1859: 338.
201. See Bailey et al. 2007.
202. See Brasier and Callow 2007; Brasier et al. 2009.
203. In one noteworthy paragraph, Darwin challenged the idea that all old rocks have been cooked and squeezed (1859: 307–8), only to imply on another page that very ancient rocks will have been hugely altered, perhaps by pressure from the deep sea (pp. 309–10, 343).
204. See, for example, Brasier and Callow 2007.
205. Darwin 1837: 505; 1881: 4.
206. See Darwin 1881: 308, 316.
207. It has been called the 'Cambrian substrate revolution' but it also involved the water column. See Bottjer et al. 2000 and Brasier et al. 2009.
208. See Logan et al. 1995. Geochemical evidence for this cleansing of the ocean shortly before the Cambrian is arguably provided by Fike et al. 2006 from Oman, and by Canfield et al. 2007 from Canada, though each of these may be only basin-wide phenomena and global evidence is awaited.

209. See, for example, McIlroy and Logan 1999, and McIlroy et al. 2003.
210. Only in unusual circumstances, therefore, such as in stagnant lakes or land-locked seas were later conditions ever able to approach those of the ancient Ediacaran and Cambrian seafloor and to allow excellent preservation to again take place.

REFERENCES

Ansted, D. T. (1866). *The Physical Geography and Geology of the County of Leicestershire*. Westminster, London.

Antcliffe, J. B., and Brasier, M. D. (2008). *Charnia* at fifty: developmental models for Ediacaran fronds. *Palaeontology*. 51, 11–26.

Bailey, J. V., Joye, S. B., Kalanetra, K. M., Flood, B. E., and Corsetti, F. A. (2007). Evidence of giant sulphur bacteria in Neoproterozoic phosphorites. *Nature*, 445, 198–201.

Bak, P. (1997). *How Nature Works. The Science of Self-organized Criticality*. Oxford University Press, 212 pp.

Ball, P. (1999). *The Self-Made Tapestry. Pattern Formation in Nature*. Oxford University Press, 287 pp.

Barghoorn, E., and Tyler, S. (1965). Microfossils from the Gunflint chert. *Science*, 147, 563–77.

Bengtson, S., and Zhao, Y. (1997). Fossilized metazoan embryos from the earliest Cambrian. *Science*, 277, 1645–8.

Bottjer, D. J., Hagadorn, J. W., and Dornbos, S. Q. (2000). The Cambrian substrate revolution. *GSA Today*, 10(9), 1–7.

Brasier, M. D. (1975). An outline history of seagrass communities. *Palaeontology*, 18, 681–702.

—— (1979). The Cambrian radiation event. In M. R. House (ed.), 'The Origin of Major Invertebrate Groups'. *Systematics Association Special Volume* 12, 103–59.

Brasier, M. D. (2000). The Cambrian Explosion and the slow burning fuse. *Science Progress, Millennium Edition*, 83, 77–92.

—— and Antcliffe, J. B. (2004). Decoding the Ediacaran Enigma. *Science*, 305, pp. 1115–17.

—— and Callow R. H. T. (2007). Changes in the patterns of phosphatic preservation across the Proterozoic-Cambrian transition. *Memoirs of the Association of Australasian Palaeontologists*, 34, 377–89.

—— and Lindsay, J. F. (1998). A billion years of environmental stability and the emergence of eukaryotes. New data from northern Australia. *Geology*, 26, 555–8.

—— Antcliffe J. B., and Callow, R. (2009). Evolutionary trends in remarkable preservation across the Ediacaran–Cambrian transition and the impact of Metzoan Mixing. In P. Allison and D. J Bottjer (eds.), *Taphonomy: Process and Bias Through Time*. Plenum Press, New York (in press).

—— Cowie, J. W., and Taylor, M. E. (1994). Decision on the Precambrian- Cambrian boundary stratotype. *Episodes*, 17, 3–8.

—— Dorjnamjaa, D. and Lindsay, J. F. (1996). The Neoproterozoic to early Cambrian in southwest Mongolia. *Geological Magazine* 133, 365–369.

—— Green, O., and Shields, G. (1997). Ediacarian sponge spicules from southwestern Mongolia and the origins of the Cambrian fauna. *Geology* 25, 303–6.

—— Rozanov, A. Yu. and Zhuravlev, A. Yu., Corfield, R. M., and Derry, L. A. (1994). A carbon isotope reference scale for the Lower Cambrian succession in Siberia: Report of IGCP Project 303. *Geological Magazine*, 131, 767–83.

Briggs, D. E. G., and Crowther, P. E. (2001). *Palaeobiology II*. Blackwell, Oxford, 583 pp.

—— Erwin, D. H., and Collier, F. J. (1994). *The Fossils of the Burgess Shale*. Smithsonian Institution Press, Washington DC, 238 pp.

Browne, J. (2003*a*). *Charles Darwin. Voyaging*. Pimlico, Random House, London, 605 pp.

—— (2003*b*). *Charles Darwin. The Power of Place*. Pimlico, Random House, London, 591 pp.

Bryson, B. (2003). *A Short History of Nearly Everything*. Black Swan, London, 686 pp.

Burkhardt, F. (ed.) (1996). *Charles Darwin's Letters. A Selection.* Cambridge University Press, 249 pp.

Buss, L. W., and Seilacher, A. (1994). The phylum Vendobionta: a sister group to the Eumetazoa. *Palaeobiology*, 20 (1), 1–4.

Canfield, D. E., Poulton, S. W., and Narbonne, G. M. (2006). Late-Neoproterozoic deep-ocean oxidation and the rise of animal life. *Science*, 315, 92–5.

Carpenter, W. B. (1891). *The Microscope and its Revelations.* J. & A. Churchill, London, 1099 pp.

Cavalier-Smith, T, Brasier, M. D., and Embly, T. M. (eds.) (2006). Major steps in cell evolution: palaeontological, molecular and cellular evidence of their timing and global effects. *Philosophical Transactions of the Royal Society, Series B*, vol. 361.

Chambers, R. (1844). *Vestiges of the Natural History of Creation.* George Routledge and Sons, 390 pp. Facsimile published by Leicester University Press in 1969.

Conan Doyle, S. (1912). *The Lost World.* Hodder & Stoughton, London.

Cloud, P. E. and Nelson, C. (1966). Phanerozoi-Cryptozoic and related transitions. *Science*, 154, 766–770.

Conway Morris, S. (1998). *The Crucible of Creation. The Burgess Shale and the Rise of Animals.* Oxford University Press, 244 pp.

—— (2003). *Life's Solution. Inevitable Humans in a Lonely Universe.* Cambridge University Press, 486 pp.

Cowie, J. W., and Brasier, M. D. (eds.) (1989). '*The Precambrian-Cambrian Boundary*'. Oxford Monographs in Geology and Geophysics, No. 12, Clarendon Press, Oxford, 209 pp.

Cribb, S., and Cribb, J. (1998). *Whisky on the Rocks. Origins of the Water of Life.* British Geological Survey, Nottingham, 72 pp.

Cross, A. B. (1928). *Charnwood Poems.* Chronicle Press, Nuneaton, 66 pp.

Cutler, A. (2004). *The Seashell on the Mountain Top.* Arrow Books, London, 228 pp.

Daly, R. A. (1907). Direct chemical evidence from truly ancient seas. The limeless ocean of pre-Cambrian time. *American Journal of Science*, 23, 93–115.

Darwin, C. (1837). On the formation of vegetable mold. *Transactions of the Geological Society, London.* Volume V, p. 505.

Darwin, C. (1845). *Journal of Researches into the Geology and Natural History of the various countries visited by H.M.S. Beagle.* 2nd edn., London.
—— (1859). *On the Origin of Species by Means of Natural Selection.* 1st edn. John Murray, London, 513 pp.
—— (1861). *On the Origin of Species.* 3rd edn. John Murray, London, 538 pp.
—— (1872). *On the Origin of Species.* 6th edn. John Murray, London, 458 pp.
—— (1881). *Vegetable Mould and Earthworms.* John Murray, London, 328 pp.
—— (1897). *Journal of Researches.* Ward Lock & Co., London, 492 pp.
Darwin, E. (1794–96). *Zoonomia, or the Laws of Organic Life.* Dublin (printed for P. Burne and W. Jones), 2 vols.
David. T. W. E., and Tillyard, R. J. (1936). *Memoir on Fossils of the Late Precambrian (Newer Proterozoic) from the Adelaide Series, South Australia.* Angus and Robertson Ltd, Sydney, Australia, 122 pp.
Dawson, J. W. (1888). *The Geological History of Plants.* Kegan Paul, Trench & Co., London, 290 pp.
Desmond, A., and Moore, J. (1992). *Darwin.* Penguin, 808 pp.
Donoghue, P. C. J., Bengtson, S., Dong, X., Gostling, N. J., Huldtgren, T., Cunningham, J. A., Yin, C., Zhao, Y., Peng, F., and Stampanoni, M. (2006). Synchrotron X-ray tomographic microscopy of fossil embryos. *Nature*, 442, 680–3.
Emsley, J. (2000). *The Shocking History of Phosphorus. A Biography of the Devil's Element.* McMillan, London, 326 pp.
Fedonkin, M. A. (1990). Systematic description of the Vendian Metazoa. In B. S. Sokolov and A. B. Iwanowski (eds.), *The Vendian System. Volume 1. Palaeontology*, Springer-Verlag, Berlin, pp. 71–120.
—— Gehling, J. G., Grey, K., Narbonne, G. M., and Vickers-Rich, P., (2007). *The Rise of Animals: Evolution and Diversification of the Kingdom Animalia.* John Hopkins University Press, Baltimore, 327 pp.
—— and Waggoner, B. M. (1997). The late Precambrian fossil *Kimberella* is a mollusc like bilatarian organism. *Nature*, 388, pp. 868–71.
Fike, D. A., Grotzinger, J. P., Pratt, L. M., and Summons, R. E. (2006). Oxidation of the Ediacaran Ocean. *Nature*, 444, 744–7.
Ford, T. (1958). *Pre-Cambrian fossils from Charnwood Forest.* Proceedings of the Yorkshire Geological Society, 31, 211–17.

—— (2007). *Charnia masoni*—50th Birthday Party. *Mercian Geologist*, 16, 280–4.

Fortey, R. (2000). *Trilobite! Eyewitness to Evolution.* HarperCollins, London, 269 pp.

Gabbott, S., Hou Xian-guang, Norry, M., and Siveter, David. (2004). Preservation of Early Cambrian animals of the Chengjiang biota. *Geology*, 32(10), 901–4.

Gehling, J. G. (1987). Earliest known echinoderm—a new Ediacaran fossil from the Pound Supergroup of South Australia, *Alcheringa*, 11, 337–45.

Germs, G. J. B. (1972). New shelly fossils from Nama Group, South West Africa. *American Journal of Science*, 272, 752–61.

Glaessner, M. F. (1984). *The Dawn of Animal Life. A Biohistorical Study.* Cambridge University Press, 244 pp.

—— and Wade, M. (1966). The late Precambrian fossils from Ediacara, South Australia. *Palaeontology*, 9, 599–628.

Goldring, R., and Curnow, C. N. (1967). The stratigraphy and facies of the late Precambrian at Ediacara, South Australia. *Journal of the Geological Society of South Australia*, 14, 195–214.

Gould, S. J. (1989). *Wonderful Life. The Burgess Shale and the Nature of History.* W. W. Norton and Co., New York, 323 pp.

Grazhdankin, D. (2004). Patterns of distribution in the Ediacaran biotas. Facies versus biogeography and evolution. *Paleobiology*, 30, 203–21.

—— and Gerdes, G. (2007). Ediacaran microbial colonies. *Lethaia*. 40, 201–210.

Grotzinger, J. P., Watters, W. A., and Knoll, A. H. (2000). Calcified metazoans in thrombolite-stromatolite reefs of the terminal Proterozoic Nama Group, Namibia. *Paleobiology*, 26, 334–59.

Guillaume, L. and Le Guyader, H. (2006). *The Tree of Life.* Éditions Belin, Paris, 560 pp.

Gürich, G. (1933). Die Kuibis-Fossilien der Nama-Formation von Sudwestafrika. *Palaontologische Zeitschrift*, 15, 137–154.

Hallam, A. (1983). *Great Geological Controversies.* Oxford University Press, 182 pp.

Harland, W. B. (1974). The Precambrian-Cambrian boundary, pp. 15–42. In C. H. Holland (ed.) *Cambrian of the British Isles, Norden and Spitsbergen.* Wiley, London, 300 pp.

Heath, P. (1974). *The Philosopher's Alice.* St Martin's Press, New York. 249 pp.

Herbert, S. (2005). *Charles Darwin, Geologist.* Cornell University Press, Ithaca and London. 485 pp.

Hockney, D. (2001). *Secret Knowledge. Rediscovering the Lost Technology of the Old Masters.* Thames and Hudson, London, 296 pp.

Hoffman, P. F., and Schrag, D. P. (2002). The Snowball Earth hypothesis: testing the limits of global change. *Terra Nova*, 14, 129–55.

Hou, Xian-Guang, Aldridge, R. J., Bergström, J., Siveter, David, J., Siveter, Derek, J., and Feng, X. H. (2004). *Cambrian fossils of Chengjiang, China.* Blackwell Scientific, Oxford, 233 pp.

House, M. R. (ed.) (1979). 'The Origin of Major Invertebrate Groups'. *Systematics Association, Special Volume* 12. Academic Press, London, 515 pp.

Huxley, T. H. (1894). *Discourses: Biological and Geological. Essays by Thomas H. Huxley* Macmillan, London, 388 pp.

Jardine, L. (2004). *The Curious Life of Robert Hooke. The Man Who Measured London.* HarperCollins, London, 422 pp.

Kemp, M. (2006). *Leonardo da Vinci. The Marvellous Works of Nature and Man.* Oxford University Press, 381 pp.

Knauth, L. P. (2005). Temperature and salinity history of the Precambrian ocean: implications for the course of microbial evolution. *Palaeogeography, Palaeoclimatology, Palaeoecology*, 219, 53–69.

Knoll, A. H. (2003). *Life on a Young Planet: The First Three Billion Years of Evolution on Earth.* Princeton University Press, 277 pp.

—— Walter, M. R., Narbonne, G. M., and Christie-Blick, N. (2004). A new period for the geologic time scale. *Science*, Washington, 305 (5684): 621.

Landing, E., Narbonne, G. M., and Myrow, P. (1988). *Trace Fossils, Small Shelly Fossils and the Precambrian-Cambrian Boundary.* Proceedings. New York State Museum Bulletin 463, 81 pp.

Leach, S., Smith, I. and Cockell, C. (2006). Conditions for the Emergence of Life on the Early Earth, *Philosophical Transactions of the Royal Society, Series B*, volume 361. Number 1474, pp. 1673–1894.

Lipps, J. H. and Signor, P. W. (eds.) (1992). *Origin and Evolution of the Metazoa.* Plenum, New York, 570 pp.

Logan, G. A., Hayes, J. M., Hieshima, G. B., and Summons, R. E. (1995). Terminal Proterozoic reorganisation of biogeochemical cycles. *Nature*, 376, 53–56.

Lyell, C. (1837). *Principles of Geology*. 5th edn. 4 vols. John Murray, London.

—— (1850). *Principles of Geology*. 8th edn. 1 vol. John Murray, London, 811 pp.

—— (1853). *Principles of Geology*. 9th edn. John Murray, London, 835 pp.

—— (1865). *Elements of Geology*. 6th edn. John Murray, London, 794 pp.

—— (1871). *The Student's Elements of Geology*. John Murray, London, 624 pp.

MacDougall, D. (2006). *Frozen Earth: The Once and Future Story of Ice Ages*. University of California Press. 267 pp.

McIlroy, D., and Logan, G. A. (1999). The impact of bioturbation on infaunal ecology and evolution during the Proterozoic–Cambrian transition. *Palaios*, 14(1), 58–72.

—— Worden, R. H., and Needham, S. J. (2003). Faeces, clay minerals and reservoir potential. *Journal of the Geological Society, London*, 160, 489–93.

—— Crimes, T. P., and Pauley, J. C. (2005). Fossils and matgrounds from the Neoproterozoic Longmyndian Supergroup, Shropshire, UK. *Geological Magazine*, 142(4), 441–55.

McIntyre, D. B., and McKirdy, A. (1997). *James Hutton, The Founder of Modern Geology*. The Stationery Office, Edinburgh, 51 pp.

McMenamin, M. (1998). *The Garden of Ediacara*. Columbia University Press, New York. 295 pp.

Marletaz, F. et al. (2006). *Chaetognath phylogenomics: a protostome with deuterostome-like development*. Current Biology, *16, R577–R578*.

Mayr, E. (2002). *What Evolution Is*. Weidenfeld and Nicolson, London, 318 pp.

Miller, H. (1858). *The Old Red Sandstone. Or New Walks in an Old Field.* Thomas Constable & Co, Edinburgh, 385 pp.

Morton, John. (2004). *King of Siluria. How Sir Roderick Murchison changed the face of Geology*. Broken Spectre Publishing, Horsham, Sussex. 280 pp.

Narbonne, G. (2004). Modular construction in the Ediacara biota. *Science*, 315, 1141–1144.

Outram, D. (1984). *Georges Cuvier, Vocation, Science and Authority in Post-Revolutionary France*. Manchester University Press, Manchester, 299 pp.

Owen, R. (1855). *Lectures on the Comparitive Anatomy and Physiology of Invertebrate Animals, Delivered at the Royal College of Surgeons*. Longman, London, 689 pp.

Parker, A. (2003). *In the Blink of an Eye: The Cause of the Most Dramatic Event in the History of Life*. Free Press, London, 316 pp.

Peach, B. N., Horne, J., Gunn, W., Clough, C. T., Hinxman, L. W., and Teall, J. J. H. (1907). *The Geological Structure of the North-West Highlands of Scotland*. Memoirs of the Geological Survey of Great Britain, 668 pp.

Peat, C. (1984). Precambrian microfossils from the Longmyndian of Shropshire. *Proceedings of the Geological Association*, 5, 17–22.

—— and Diver, W. (1982). First signs of life on Earth. *New Scientist*, 95, 776–8.

Pflüg, H. D. (1972). Systematik der jung-präkambrischen Petalonamae. *Paläontologische Zeitschrift*, 46, 5667.

Qian, Y., & Bengtson, S. (1989). Palaeontology and Biostratigraphy of the Early Cambrian Meishucunian Stage in Yunnan Province, South China. *Fossils & Strata* 24, 1–156.

Raup, D. M. (1966). Geometric analysis of shell coiling: general problems. *Journal of Paleontology*, 40, 1178–90.

Rozanov, A. Yu., et al. (1969). *The Tommotian Stage and the Cambrian Lower Boundary Problem*. Transactions of the Geological Institute, Moscow. [In Russian. An English translation was published by the US Department of the Interior in 1981.]

Runnegar, B. (1982). Oxygen requirements, biology and phylogenetic significance of the late Precambrian worm *Dickinsonia*, and the evolution of the burrowing habit. *Alcheringa*, 6, 223–39.

Rupke. N. A. (1994). *Richard Owen. Victorian Naturalist*. Yale University Press, New Haven and London. 462 pp.

Salter, J. W. (1856). On fossil remains in the Cambrian rocks of the Longmynd and North Wales. *Quarterly Journal of the Geological Society*, 12, 246–51.

—— (1857). On annelide burrows and surface markings from the Cambrian rocks of the Longmynd. *Quarterly Journal of the Geological Society*, 13, 199–207.

Schopf, J. W. (1999). *The Cradle of Life*. Princeton University Press, 367 pp.

Secord, J. (1986). *Controversy in Victorian Geology. The Cambrian-Silurian Dispute*. Princeton University Press.

Seilacher, A. (1992). Vendobionta and Psammocorallia: lost constructions of Precambrian evolution. *Journal of the Geological Society of London*, 149, 607–13.

Selden, P., and Nudds, J. (2004). *Evolution of Fossil Ecosystems*. Manson Publishing, London, 160 pp.

Sollas, W. J. (1905). *The Age of the Earth and Other Geological Studies*. T. Fisher Unwin, London, 328 pp.

Spencer, F. (1990). *Piltdown. A Scientific Forgery*. Oxford University Press, 272 pp.

Sprigg, R. C. (1947). Early Cambrian (?) jellyfishes from the Flinders Ranges South Australia. *Transactions of the Royal Society of South Australia*, 71, 212–24.

—— (1949), Early Cambrian 'jellyfishes' of Ediacara, South Australia and Mount John Kimberley District, Western Australia. *Transactions of the Royal Society of South Australia*, 73, 72–99.

—— (1989). *Geology is Fun*. Reg Sprigg, Adelaide, 347 pp.

Stewart, I. (2001). *What Shape is a Snowflake?* Weidenfeld & Nicolson, London, 223 pp.

Temple, R. (2000). *The Crystal Sun. Rediscovering a Lost Technology of the Ancient World*. Arrow Books, London, 642 pp.

Twemlow, G. (1868). *Facts and Fossils Adduced to Prove the Deluge of Noah and Modify the Transmutation System of Darwin with Some Notices Regarding Indus Flint Cores*. Simpkin, Marshall & Co., Guildford, 256 pp.

Voronin, Yu. I., and others. (1982). The Precambrian-Cambrian boundary in the geosynclinal areas (the reference section of Salany-Gol, MPR.) *Transactions of the Joint Soviet-Mongolian Palaeontological Expedition*, 18. Nauka, Moscow. [In Russian].

Walcott, C. D. (1899). Precambrian fossiliferous formations. *Bulletin of the Geological Society of America*, 10, pp. 199–244.

Walker, G. (2004). *Snowball Earth*. Bloomsbury, London, 288 pp.

Watts, W. W. (1947). *The Geology of the Ancient Rocks of Charnwood Forest*. Backus, Leicester, 160 pp.

Weinberg, S. H. (1999). *A Fish Caught in Time. The Search for the Coelacanth*. Fourth Estate, London, 239 pp.

Xiao, S. Yu Zhang, and Knoll, H. (1998). Three-dimensional preservation of algae and animal embryos in a Neoproterozoic phosphorite. *Nature*, 391, 553–8.

—— —— (1999). Fossil preservation in the Neoproterozoic Doushantuo phosphorite lagerstätte, South China. *Lethaia*, 32, 219–40.

INDEX